THE MASTER ARCHITECT SERIES
ARUP ASSOCIATES
Selected and Works

世界建筑大师优秀作品集锦
阿鲁普联合事务所

王 剑 译

中国建筑工业出版社

著作权合同登记图字：01－2003－0647号

图书在版编目（CIP）数据

阿鲁普联合事务所／澳大利亚 Images 出版集团编；王剑译．—北京：中国建筑工业出版社，2004
（世界建筑大师优秀作品集锦）
ISBN 7－112－06696－4

Ⅰ．阿... Ⅱ．①澳...②王... Ⅲ．建筑设计－作品集－美国－现代 Ⅳ．TU206

中国版本图书馆 CIP 数据核字（2004）第 058088 号

Copyright © The Images Publishing Group Pty Ltd
All rights reserved. Apart from any fair dealing for the purposes of private study, research, criticism or review as permitted under the Copyright Act, no part of this publication may be reproduced, stored in a retrieval system or transmitted in any form by any means, electronic, mechanical, photocopying, recording or otherwise, without the written permission of the publisher. and the Chinese version of the books are solely distributed by China Architecture & Building Press.

本套图书由澳大利亚 Images 出版集团公司授权翻译出版

责任编辑：程素荣
责任设计：郑秋菊
责任校对：赵明霞

世界建筑大师优秀作品集锦
阿鲁普联合事务所
王　剑　译

中国建筑工业出版社出版、发行（北京西郊百万庄）
新　华　书　店　经　销
北京嘉泰利德公司制版
恒美印务有限公司印刷
＊
开本：787×1092毫米　1/10　印张：25⅗　字数：600千字
2005年1月第一版　　2005年1月第一次印刷
定价：218.00元
ISBN 7－112－06696－4
　　TU·5850（12650）
版权所有　翻印必究
如有印装质量问题，可寄本社退换
（邮政编码 100037）
本社网址：http://www.china-abp.com.cn
网上书店：http://www.china-building.com.cn

Contents 目　　录

9　导言
　　阿鲁普联合事务所—五位专家的评述

作品精选

20 世纪 60 年代
22　伯明翰大学采矿和冶金学院
26　莫尔丁斯音乐厅
30　IBM（英国）计算机制造装配中心

20 世纪 70 年代
38　地平线项目，约翰·普莱耶和后裔的工厂
42　IBM 公司北港工程
48　亨利·伍德音乐厅
50　IBM 公司大楼，约翰内斯堡
52　布什·莱恩大楼
56　盖特威 1 号楼
62　杜鲁门有限公司总部
70　托马斯·怀特爵士大楼，牛津圣约翰学院
74　伦敦劳埃德保险公司总部大楼
78　英国中央电力局西南区总部

20 世纪 80 年代
84　特房伯有限公司工厂
88　贝德福德学校
92　巴伯夫区自治会办公楼
98　盖特威 2 号楼
104　布莱克里夫大楼
108　国际园艺节展厅
114　芬斯伯里大道 1 号
120　外交使馆区运动俱乐部
122　福布斯·梅隆图书馆，剑桥克莱尔学院
126　布鲁德门
128　主裤广场
132　斯托克利园区
136　法人—通用保险公司大楼
146　哈斯伯罗—布拉德利（英国）有限公司总部
148　帝国战争博物馆一期工程
156　斯托克利园区会所

20 世纪 90 年代
160　苏塞克斯正面看台，古德伍德赛马场
162　温特沃斯高尔夫球俱乐部
166　伯格塞尔露天赛场
168　劳埃德银行股份公司办公大楼
176　科坡斯恩酒店
178　白金汉宫大道 123 号
182　霍莎姆园区，盖莎根
184　朱比利地铁线附属服务控制中心
186　皇家保险大楼
190　柏林 2000 年奥运会场馆
192　古维尔和卡尔斯学院，剑桥
196　香港大屿山 – 机场铁路中央车站
198　幼儿园，罗森海姆，法兰克福
200　杜塞尔多夫塔楼
204　废料利用发电厂
206　里昂阿尔卑斯科技园区
208　伊斯坦布尔文化和会议中心
212　格罗大楼，新城
214　曼彻斯特 2000 奥林匹克国际体育场
218　约翰内斯堡田径运动场

事务所简介
222　个人简历
228　建筑及项目年表
236　获奖及参展作品
245　参考文献
253　致谢
254　索引

Introduction

Introduction

Arup Associates — Five Perspectives

导　　言

阿鲁普联合事务所—五位专家的评述

我们所建造的作品应该是一个整体，一个统一体，而设计工作很大程度上就是要赋予它作为艺术作品的整体性和完美性。

奥夫·阿鲁普强调整合所有设计相关决策使之一体化的必要性，从而为设计工作开创了独特的视角和方法。他倡导配合默契的小型设计团队，在共同的空间中连贯地开展工作。他相信这样的工作模式可以使得团队成员相互学习和汲取彼此的特长，在这样的条件下，所谓的领导力问题几乎不存在，每一个人都能领导着自己的方向……甚至所谓的专业界限也将消失。

这就是阿鲁普联合事务所1963年成立之际所秉承的原则。随着建筑物变得越来越复杂多变，工程建设者需要了解更全面更广泛的信息，不仅仅是设计事务，还要包括规划、建造、法律、工程技术、环境景观、成本控制和合同管理。因此，随着工程实践的需要，以多学科交叉的小型团队为主体的工作组织模式得到发展。这种团队包括建筑师、结构工程师、环境工程师、预算师和室内设计师等等。设计团队可以在从前期规划、初步设计的形成一直到工程资料的准备和最终决算的全过程保持与用户的直接沟通。

这样的设计工作模式成功地迎合了环境和人类的需要，同时也融合了技术进步带来的益处，本书中详尽介绍的这些工程项目就是实例。

工业建筑、实验室建筑和科研设施的设计经验使阿鲁普联合事务所相信在结构的整体性设计和建造过程自有的理性次序有特别的关联。这种关联在早期的工程如CIBA有限公司和埃沃德有限公司的工厂项目中得到发展，在后来的伯明翰大学采矿和冶金学院工程的设计中得到升华——这一项目采用垂直相交的单元，使设备管线可以清晰连贯地布置在结构板层中。这样的设计来源于对建筑功能的独到见解，来源于对结构设计、管线布置和施工技术这三者的系统把握。

What we build should always be a whole, an entity, and the job of designing it is very much the job of giving it the wholeness of a work of art and the inevitability of the perfect tool.

In emphasising the need to integrate all of the design decisions relating to a project, Ove Arup inspired a very particular view of design and a way of working. He advocated small, closely knit teams of designers, all working in the same space and having a continuity of work on a few jobs at a time. It was a way of working that he felt could enable the team to learn from and appreciate each other's unique qualities and where "the question of leadership need hardly arise, each member taking the lead in his own subject . . . even the professional demarcations may fall away".

It was this philosophy that was enthusiastically adopted as the basis for the foundation of Arup Associates in 1963. As buildings have become more complex and varied, so those who commission them need more comprehensive advice, not only in matters of design, but also in planning, construction, law, engineering, environmental servicing, cost control and contract management. Consequently, the practice has developed the skills and organisation to tackle design projects within a framework of small, multi-disciplinary teams that include architects, structural and environmental engineers, cost estimators and interior designers. Those design teams work directly with the client from the development of the brief and formation of initial design responses to the preparation of production information and the settlement of the final account.

This is a way of working which, as the projects illustrated in this book so clearly demonstrate, has successfully developed ideas that are sensitive to the environment and human needs, yet which also obviously profit from technological innovation.

Experience in the design of industrial buildings, laboratories and research facilities encouraged a particular concern within the practice for the integration of structure and the rational order established by the building process itself. These concerns, developed in early projects such as those for CIBA and Evode, were subsequently refined in the design for the Department of Mining and Metallurgy at the University of Birmingham. This design was planned on a tartan grid within which routes for the distribution of services were clearly designated between a series of structural tables. The form of this project was based on a framework that grew out of detailed studies of building uses and highly original interpretations of the systems of structure, servicing and construction.

Introduction Continued

阿鲁普联合事务所随后的作品也都建立在这样的基础上，在教育机构和其他类型建筑的设计中继续努力地改进工作空间的布局，使这种系统化思想得以实现和发展。

20世纪60年代晚期，阿鲁普联合事务所参与了拉夫伯勒大学的规划和其中一些单体建筑的设计，另外，还承担了诺丁汉的地平线项目的设计，这些项目的设计进一步发掘了正交建筑单元的潜力。设计师们把这些建筑分割成一系列的样图和模型，精确细致地标定和描绘建筑的部件和拼接。新式的施工技术与设计得到了密切的结合，保证这些规模巨大而复杂的工程在短时间内以得以高水平的设计与完成。

在设计新的市政、公司、办公建筑的过程中，设计师们也开始考虑如何将理性的系统化思想和由当地环境、城市架构等因素决定的个性化需要统一起来。公司总部建筑的设计，例如杜鲁门有限公司、英国中央电力局（CEGB）、伦敦劳埃德保险公司查莎姆总部、威金斯·特普、IBM英国有限公司等等，都创造性地消除了体系和环境之间的原本显而易见的冲突。

经过15年的发展，这些得益于我们多学科团队协作工作模式的设计作品，转变了人们对办公工作空间和建筑形态的看法。在这些项目中，建筑结构不仅仅是定义室内外空间，还要与周边环境相融合。场地规划和定位的概念得到发展，并融合在建筑外观和构造细节之中。因此，在杜鲁门有限公司、英国中央电力局、伦敦劳埃德保险公司、盖特威1号楼等项目中，可以看到结构开间和建筑构造是开放的，可以清晰地分隔使用空间和设备管线区域。同时，杜鲁门有限公司和伦敦劳埃德保险公司查莎姆总部的设计与城市的轮廓相协调，英国中央电力局、盖特威1号楼则改善了城郊的自然景观。

The subsequent work of Arup Associates built on this remarkable foundation. Designs for educational institutions and for other clients keen to improve the design of the workplace, enabled those systemic approaches to be developed and refined.

During the late 1960s, the master plan for a new university at Loughborough, followed by proposals for new buildings there, and for the Horizon Project in Nottingham, further explored the potential of the tartan grid. The designers dissected those buildings in a series of drawings and models, identifying the components and plotting their assembly with painstaking precision. This unusual concern for the making of buildings, combined with innovative construction techniques, made these large and complex projects realisable in a short time and to unusually high standards of design and finish.

The problems of designing new corporate offices encouraged the same designers to consider how best to realise the benefits of this rational systemic approach while at the same time responding to the particular requirements of sites within sensitive natural landscapes or the dense fabric of cities. The design of new headquarters buildings for corporate clients, such as those for Truman Ltd, CEGB, Lloyd's of London at Chatham, Wiggins Teape or IBM, demonstrated highly original resolutions to the often apparently conflicting demands of system and setting.

Developed over a 15 year period, the designs of these projects, clearly inspired by the collaborative efforts of the multi-disciplinary team, transformed attitudes about the nature of office workspace, building form and envelope. In these projects the structure of the building not only defines inside space and outdoor room, but has also been designed to incorporate concepts of environmental design. Ideas about site planning and orientation have been developed to inform both building configuration and tectonic detail. So at Truman Ltd, Lloyd's of London, CEGB and Gateway 1, the structural bay and building fabric are exposed so as to clearly define places to work, to create spaces for services and to be thermally responsive. At Truman Ltd and Lloyd's of London at Chatham, these systems have been ordered to reveal traces of the city and sensitively repair urban sites; the designs for the CEGB and Gateway 1 reinstated natural landscapes on the suburban edge.

这些对解决体系和环境之间冲突的尝试影响了设计规划，影响了与室内和户外空间的布置，影响了材料选择和外部装饰设计。这些来源于实践的想法尝试性而又周密地应用在后来的一些作品中，例如，勒瑟和戈德温项目，Legal & General 规划，皇家人寿保险公司和劳埃德银行项目等。

阿鲁普联合事务所在保护修复古建筑和建造演出场所方面做了许多贡献，例如，第一个作品就是为作曲家本杰明·布里顿建造的莫尔丁斯音乐厅。另外还有亨利·伍德音乐厅、格拉斯哥苏格兰歌剧院和布克斯顿歌剧院等等，都是将原有的古建筑修复投入使用，这些为数众多的重要工程，都体现了兼顾现代技术需要和古建筑保护的设计技巧。在东英吉利亚大学，阿鲁普联合事务所承担了音乐学院新建筑的设计。这些设计技巧在更晚一些的作品中得到进一步发展，如伦敦大英帝国战争博物馆的分期开发，在竞赛中获奖的伊斯坦布尔新文化中心的设计方案等。

阿鲁普联合事务所承担过许多大型的工程开发项目，例如，在伦敦市中心一块 $3.2hm^2$ 的土地上设计新的大型金融中心，在伦敦机场附近一块约 $140hm^2$ 的荒地中建造新的国际商业区等等。这些大型工程越来越要求设计团队在原有的多学科背景的基础上，进一步拓展专业范畴和视野。地质工程师、交通规划师、考古学家、景观建筑师和建造专家都被吸纳到设计团队中，为城市设计发展了一系列新概念。

这样的设计思想也对欧洲一些大型工程项目开发产生了深远的影响。在法国、匈牙利和德国的一些新的商业区的设计方案中，都考虑了场地、景观、建筑设计的贯通，创造性地将公共和私人用途一体化。

These explorations of the areas of conflict between system and setting have influenced the geometries of plan; the formation of outdoor spaces to link the wider context with the indoor room; the materials used; and the design of responsive external skins. The ideas contained within the designs of these projects were tested in use and thoughtfully developed in subsequent schemes for Leslie & Godwin, Legal & General, Royal Life and Lloyds Bank.

At the same time, Arup Associates were making an outstanding contribution to the restoration of historic buildings and the creation of new spaces for music and performance. The Maltings Concert Hall, designed for Benjamin Britten, was the first of many important projects that established within the practice skills to create design that combined modern uses with the restoration of historic buildings. Designs for the Henry Wood Hall, the Scottish Opera in Glasgow and the Buxton Opera House all restored historic buildings and returned them to use, while the Music School added new buildings to the campus of the University of East Anglia. More recently these skills have been extended in the designs for the phased long-term development of the Imperial War Museum in London and the competition-winning scheme for the new Cultural Centre in Istanbul.

Commissions to design a large new financial centre at Broadgate on a 3.2-hectare site in the heart of the City of London, and the creation of a new international business community at Stockley Park on 140 hectares of contaminated land near London Airport, generated an increasing interest in gathering a broader range of skills in the multi-disciplinary team. With the collaboration of geotechnical engineers, transportation planners, archaeologists, landscape architects and construction specialists, new ideas and concepts in urban design have been developed.

This experience has led to an increasing involvement in the design of large and complex projects, several in Europe. Proposals commissioned for the design of new business communities in France, Hungary and Germany integrate public and private uses in innovative plans for site, landscape and buildings.

Introduction Continued

在早期的利物浦、利雅得、古德伍德、温特沃斯等地的体育建筑项目中取得成功后，近年来我们也致力于建造新的运动场馆。在柏林和曼彻斯特，这些大型建筑项目被当作是城市大规模改造的催化剂，它们用于举办国际的体育赛事。这些项目的建造体现了开发商和建造商的密切合作，他们都设想了巧妙的方案，使得这些新的体育建筑能够在城市的改造中发挥独到的作用。南非约翰内斯堡为1995年世界杯橄榄球赛所建造的城市体育场就使得市区的重要地段焕然一新。

以上罗列的建筑设计和规划范畴广泛，都很好地体现了我们的设计师的技巧和他们多学科协作工作模式的优越。奥夫·阿鲁普将建筑定义为"使人们心灵愉悦的营造手段"。正是这种思想持续促进着阿鲁普联合事务所设计人员与客户之间的协作和沟通，而客户给我们的评价也最好地体现了这种协作。

布赖恩·卡特
阿鲁普联合事务所

Following the success of earlier projects for sports buildings at Liverpool, Riyadh, Goodwood and Wentworth, the practice has also recently developed proposals for new stadia. Planned to act as catalysts for the larger scale urban regeneration of extensive sites in Berlin and Manchester, these projects were prompted by plans to host major international sporting events and were developed in close collaboration with developers and builders. They each outline ingenious ways of integrating large new sports buildings with a mix of other uses to reconstruct the city. A new urban stadium designed to provide facilities for the 1995 Rugby World Cup while also regenerating an important segment of the city, is currently under construction in Johannesburg.

This broad range of outstanding work in the design of building and city emphasises the skills of these particular designers and the significance of their multi-disciplinary approach. Ove Arup defined architecture as "a way of building which delights the heart". It is that vision which clearly continues to inspire the commitment and collaboration of designer and client at Arup Associates. The responses of our clients that follow vividly describe the nature of those collaborations.

Brian Carter
Arup Associates

斯坦霍普房地产开发公司（Stanhope Properties PLC）

阿鲁普联合事务所将品质视为设计作品的生命。但这究竟意味着什么呢？怎么样才算高品质的设计？不同的人会有不同的看法。对于一些人来说，这纯粹取决于建筑的立面；而对另外一些人，他们可能关心的是建筑物内部的人体工程学细节，或者是建筑物使用的效能，或仅仅是大理石和花岗石的数量。从一个专业开发商的角度来说，他关心的是价值和价格的关系，用户需要、美学、投资者期望和建筑功能之间的平衡。对于开发商，高品质设计就是将这些通常冲突的需求、约束和影响协调和一体化。

阿鲁普联合事务所采用的多学科协作的团队工作模式，使得涉及广泛范畴的不同学科能够非同寻常地默契配合。他们的工作方式——整个团队在共同的空间中工作，团队内部经常性的互相批评和热烈讨论，特别有助于开展创新性的工作。他们处理问题的方式往往大大领先于潮流，例如更有效地使用材料和能源。这些团队积累的经验和信心使他们能够很快地对复杂问题做出可靠的判断。他们的作品看似简单，但由于将结构工程和环境工程的设计与建筑学融合成了一体，往往能一下子解决一大串问题。

阿鲁普联合事务所的作品的另一个与众不同之处是他们设计中体现出的人性化。他们竭力避免使建筑仅仅成为一个符号，而是努力使城市具有更加持续的品质——这正是战后的英国大多数城市所不幸缺少的。斯托克利园区和布鲁德门就是很好的例子，在伦敦的修复和重建中它们是两个主要部分。新颖的总规划起步于对户外空间的设计，然后才是对实体建筑的设计，从而使得内部和外部空间协调，形成一体化的环境。

Introduction Continued

这样复杂的工程项目对设计者以及所有相关的人来说都会是一个严峻的考验。很好的例子就是现在被认为是欧洲领先的斯托克利商业园区。阿鲁普联合事务所的多学科协作的团队工作模式在这样的情况下大放异彩。建筑学、规划、景观设计、征用土地、污染控制、土木工程等等学科，在同一个对商业区概念的重新诠释之下被联合在一起。这保证所有项目在非常严格的预算和时间约束下完成。没有他们的技能、智慧、想像力和专业水准，斯托克利园区项目是不能取得今天这样的成功的。

卓越的结果只能通过在所有日益复杂和漫长的研究、规划、设计和建造过程中不断追求完美得到。高品质不是随意可得，也不可能永久保证。它和技巧、理念和管理有关。阿鲁普联合事务所的作品一贯体现了这种理念。他们的设计提供了美观和恰到好处的环境，促进了人与人之间的交流，同时又符合所有的技术要求：这些都是了不起的成就。

文森特·王（Vincent Wang）
斯坦霍普房地产开发公司董事

Such complex projects test to the limit the competence of all involved and provide the acid tests of excellence in the design process. Nowhere is this more apparent than at Stockley Park, now widely acclaimed as Europe's leading business park. Arup Associates' multi-disciplinary approach shines under such circumstances. Architecture, master planning, landscape design, land reclamation, pollution control and civil engineering had to be combined with an understanding of what was then a new property concept — the business park — all to be delivered against a very challenging budget and timescale. Without their skills, ingenuity, imagination and sheer professionalism Stockley Park would not have become the success it is today.

Excellence in the end result can only be achieved by excellence in all the increasingly complex and lengthy processes of research, planning, design and construction. Quality is not an optional extra, and can never be bolted on afterwards. It is, rather, a question of skill, commitment and loving care. Arup Associates' work consistently manifests an understanding of this concept. Their designs provide beautiful and appropriate environments and promote good personal interaction, at the same time meeting all the requisite technical demands: these are huge achievements.

Vincent Wang
Director, Stanhope Properties PLC

IBM（英国）有限公司

从我们在英国开展业务的初期起，我们就一直和阿鲁普联合事务所保持着合作关系。而同时，阿鲁普联合事务所也在英国逐步成长为建筑业的知名品牌。

追溯过去，曾经有一段时期，IBM取得了空前的增长，为了适应日益扩大的业务规模，我们计划在英国建造新的制造基地，还计划寻找一个合适的地方建造公司总部以代替当时在伦敦市中心租赁的写字楼。

我们在英格兰的南部买下了两块大面积的土地，以满足这两个需求。一块在哈万特，后来建造了我们庞大的制造工厂；另一个地方则建造了公司的英国总部，可以容纳3000人办公。在它们的建造过程中，我们一直坚持和我们认为有着创新精神和独特活力的团队保持协作，从而顺利地完成了这些工程的设计和建造。

为了说明这种协作的重要性，我必须强调商业公司在这种大规模建设项目的运作中可能面临的种种风险。巨大的花费都来源于公司过去积累的利润，而我们必须委托外部的公司贯彻我们对建筑的理解和期望。

我们也不能低估问题的复杂性。举个例子来说，有一些复杂的生产车间，他们的生产任务经常会改变，因此就要求空间开阔且可以随时变换设置，以适应新的生产流程。另一个例子是，公司总部在其建筑的使用寿命期内，容纳的办公人员的数量和要求都可能发生很大的变化。

位于哈万特和北港的这两块土地本身的特点也需要特别的解决方案，它涉及到十分复杂的场地土壤条件、土地的征用、交通、周边环境和对景观的重新规划等等。工程师必须和建筑师携手合作解决这些问题。

Introduction Continued

除此之外，我们还希望我们的场所能够得到可持续的发展，建筑可以适应不断变化的要求，空间适于工作而且有很高的利用效率。阿鲁普联合事务所和我们一起，分析和完成可行性方案、总体规划和具体设计，使我们的期望得以顺利实现。

我们位于北港的公司总部使用的土地是填海而成的，阿鲁普联合事务所采用了先进而复杂的荷兰技术填筑出这片土地，使它达到开发要求。而今天我们漫步在漂亮的建筑和景观之中时，也许根本就意识不到当初在这片土地上所付出的巨大艰辛。这就是对我们的开发伙伴阿鲁普联合事务所的智慧的最好赞誉。

哈万特和北港项目都获得了奖项。在这些建筑里工作的人们、参观和访问过它们的人尤其是我们的顾客，都对它们十分赞赏，因为这些建筑给他们创造了舒适的环境。这样的成绩应归功于建设过程中的坚韧不拔和不断创新，特别是建筑师、工程师和其他学科人员相互协作的团队精神。作为客户，我们对此结果也十分满意。

彼得·温格瑞夫（Peter Wingrave）
（英国皇家建筑师协会认证，英国皇家艺术学会荣誉会员）
设计和营造经理
IBM（英国）有限公司

More than that, however, we wanted sites where development could grow logically; buildings that could respond to ever-changing requirements; and spaces that would be both efficient and a delight to work in. Arup Associates worked with us to analyse and devise feasibility plans, master plans and designs that would allow us to grow and expand logically without detriment to everything that had gone before.

Our headquarters site at North Harbour was reclaimed from the sea. Incredibly complex Dutch techniques were adopted by Arup Associates to reclaim the land before it could be developed. Today we are unaware of this tremendous effort when we stroll around the beautifully landscaped site. This is a credit to the ingenuity of our partners in this development.

The projects at Havant and North Harbour have won awards, and deservedly so; they have also won acclaim from all the thousands of people who have had the pleasure to work there, or pass through, in particular our many customers. All of this is due to the tenacity and ingenuity of the practice who conceived and created them, and in particular the tremendous teamwork between architects and engineers of all specialist disciplines and, of course, ourselves, the client. We commend the result.

Peter Wingrave (RIBA Dip Hons FRSA)
Design & Construction Manager
IBM (UK) Limited

大英帝国战争博物馆

我1982年来到帝国战争博物馆时,博物馆的建筑项目正由公共资产机构管理。他们已经指定了一个建筑师事务所作为项目的顾问,但我并不满意这个安排,而想重新委托别的事务所。我们非正式地和许多建筑师进行了接触,最后我们都认为阿鲁普联合事务所是解决我们特殊需要的最合适选择。

他们在古建筑方面已经取得的许多成绩深深吸引了我,例如他们为萨福克郡巴伯夫(Babergh)区自治会设计的新办公楼。又如阿鲁普联合事务所为利物浦园艺节设计的展馆,在我看来是一个非常有意思的结构,而且和我们所面临的问题也有相通之处。

对于帝国战争博物馆这样的建筑物,由于四周的外墙给空间造成了限制,而引发了公共空间和日益拥挤的馆藏之间的冲突。我们初步的设想是把走廊塞得更满,使用两倍高度的艺廊。是阿鲁普联合事务所对这一方案提出异议,他们认为这会破坏空间的观感,最后我们也确信了这一点。他们的团队精神和提供的优质服务都给我们留下了深刻的印象。

阿兰·博格(Alan Borg)博士
大英帝国战争博物馆馆长
(节选自作者与蒂姆·奥斯特勒的谈话,原载于《世界建筑》1992年第17期)

Imperial War Museum

When I arrived at the Imperial War Museum in 1982, museum building projects were managed for us by the Property Services Agency. They had appointed an architectural practice as consultants, but I wasn't happy with this arrangement and wanted to commission a new practice. We talked informally to a number of architects, but I think we all felt that Arup Associates were going to be the right sort of architects for the particular problems we had.

I was impressed by the various things they had done with historic buildings, as well as the new offices they had done for the Babergh District Council in Suffolk. I also looked at the pavilion Arup Associates had designed for the Liverpool Garden Festival, which seemed to me an interesting structure, and one which dealt with some of the same problems that faced us.

In a building such as the Imperial War Museum there tends to be a conflict because the space is limited by the outside perimeter wall. It is the problem of open space versus crowded contents. Our initial proposal was to have the aisles much fuller, with a double height art gallery. It was Arup Associates who thought that the space would look much better without that, and we were eventually convinced. We were especially impressed by their team approach and the amount of consultation they provided.

Dr Alan Borg
Director, Imperial War Museum

(Extract from an interview with Tim Ostler first published in World Architecture, no. 17, 1992.)

Introduction Continued

剑桥克莱尔（Clare）学院

17世纪末，克莱尔学院建造了剑桥第一个现代意义的图书馆。在这之前，书籍都堆在大木箱里，放在古老的教堂的阁楼里。新图书馆工程建造在重建后的老广场的北侧。我们不能确定这个图书馆是何时投入使用的，不过在1742年剑桥的文物学家威廉·科尔这样写道：

这是大学中超一流的图书馆，藏书室十分宽敞且比例得当，富有现代气息，书籍就排列在四周，而不是像其他学院图书馆那样堆在箱子里……

在房间的中间留有着充裕且舒适的工作空间也是当时设计的优点。

然而，自19世纪末之后，克莱尔学院渐渐地落伍了。馆藏不断地扩大，但是利用率却不成比例。它从不对本科生开放。1926年曼斯菲尔德·福布斯在参加学院成立600年庆典时，褒扬了老图书馆的藏书广泛，但也不得不为学院的书籍服务方面的失败向学生和公众致歉。

1935年福布斯逝世，他将个人藏书全部捐献给学院，建立了一座为本科生服务的图书馆。在他局促的住所，福布斯图书馆为克莱尔学院的学生服务了30年。但是到了1970年代中期，它的能力也到了极限。

到了这个时候，学院的需要已经不仅仅是扩大图书馆的规模了。二战之后学院也发生了很多变化，1966年克莱尔的礼堂落成，此外最重要的变化还有自1972年开始招收女生。学院的唱诗班也随着变成男女混合。同时，由于有三位天才般的音乐教授，大量的歌唱家和演奏家为之吸引，进入学院深造。这使得克莱尔学院在音乐方面异军突起，具有显著的竞争力。

1979—1980 年，学院审时度势，认为当前最紧迫的需要包括建造一个新的本科生图书馆，为音乐家们建造新的设施，以及为越来越多的通过学位考试的研究生们提供设施。1983 年秋天，学院启动了 Thirkill-Ashby 计划，筹款 125 万英镑，用于扩建福布斯图书馆，一个演奏厅和音乐练习室。

建筑师们提出的初步规划中，试图把图书馆和音乐场所设计在同一建筑之中，场地就选在古老的纪念广场旁。建筑的风格和功能都会很现代化，但同时也和原来的古建筑保持一致。主管部门接受了阿鲁普联合事务所的设计，建筑在 1986 年复活节前夕正式对学生开放。

虽然场地狭小，但新建筑却并没有局促的感觉，它不仅包含了新的图书馆和隔音良好的音乐设施，还有一个公共休息室，一个影印室和一个计算机房。建筑坐落在古老的圣伊莱斯·吉尔伯特·斯科特广场（Giles Gibert Scott's orginal court）的中央，由于很好地考虑到与周边建筑的关系，使得它并没有打破斯科特的设计格局，而相反是使其更完美。建筑中心的八角形门厅可以使人追溯到 18 世纪时老教堂的八角形前庭。和老图书馆一样，新的图书馆中书籍也挨着外墙布置，中央留出充裕而舒适的工作区域。

这一建筑是学院发展道路上的一个新的里程碑。斯科特的设计的两大缺陷都被消除，原有建筑之间的关系更为协调，而且都能发挥恰当的作用。

理查德·古德（Richard Gooder）博士
福布斯图书馆馆长，剑桥大学克莱尔学院校务委员会委员

Reviewing its position in the academic year 1979-80, the College concluded that the most pressing needs were for a new undergraduate library, facilities for musicians and support for the research students whose numbers were growing in consequence of Clare's tripos success. In the autumn of 1983 the College launched the Thirkill-Ashby Appeal, with the object of raising £1.25 million for the purpose of housing an enlarged Forbes Library, a recital room and music practice rooms.

The architect's brief was to submit a design for the accommodation of the library and music facilities on the Memorial Court site, preferably within a single building. The building was to be as modern in style and function as consistency with the existing architecture of Memorial and Thirkill Courts would allow. The Governing Body accepted the design of Arup Associates, and the building was opened for the use of students at the beginning of Easter Term in 1986.

On a small site the architects have included, with no sense of constriction, not only the library and properly insulated music facilities, but a common room, a photocopying room and a computer room as well. The building stands in the centre of Giles Gilbert Scott's original court, but with such respect for the surrounding architecture that it seems not to intrude on Scott's design, but to complete it. The octagon at the centre of the building is reminiscent of the beautiful 18th century octagonal antechapel in Old Court. As in the Old Fellows' Library, the books in the new library are ranged back against the outside walls, leaving a spacious and well-lit working area at the centre.

The building marks a new development in the life of the College. At a stroke the two main criticisms of Scott's design have been removed: the domestic scale and function of Memorial Court no longer serve merely as a triumphal approach to the University Library; and Memorial and Thirkill Courts no longer seem dormitory suburbs of Old Court.

Dr Richard Gooder
Forbes Librarian and Fellow of Clare College, Cambridge

Selected and Current Works

作品精选

1960s

Mining and Metallurgy, University of Birmingham

Design/Completion 1962/1966
Birmingham
Birmingham University
14,500 square metres
Precast concrete
Concrete and glass

This scheme for new laboratories and training facilities was planned as five blocks, giving each department a separate building.

As future patterns of growth were unpredictable, the design was developed to allow for growth in any direction, with all services available at any point. The structural unit is a square tower, spanned by a single coffered slab supported at each corner by a cluster of four tied columns. Services run vertically at the centre of the column clusters, and horizontally in the 1 metre depth between adjoining slabs. Electrical trunking runs in grooves under the coffer ribs.

The towers are massed into squares five or six bays wide, overlapping one bay at the junctions. Each working bay of the building is surrounded by a border

Continued

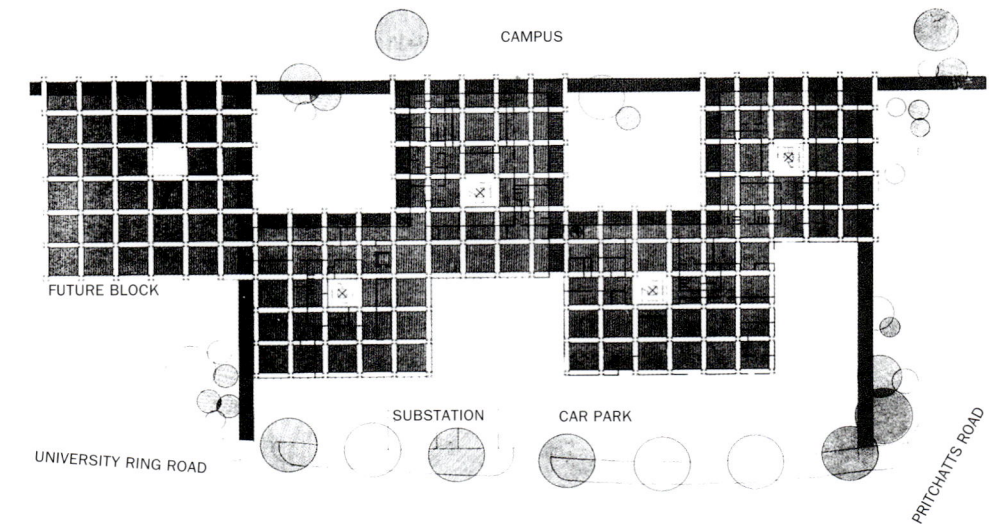

1. Site plan showing the five linked blocks
2. The colonnade; the column cluster is capped by fume cupboard extract hoods
3. Perspective section showing the elements of construction

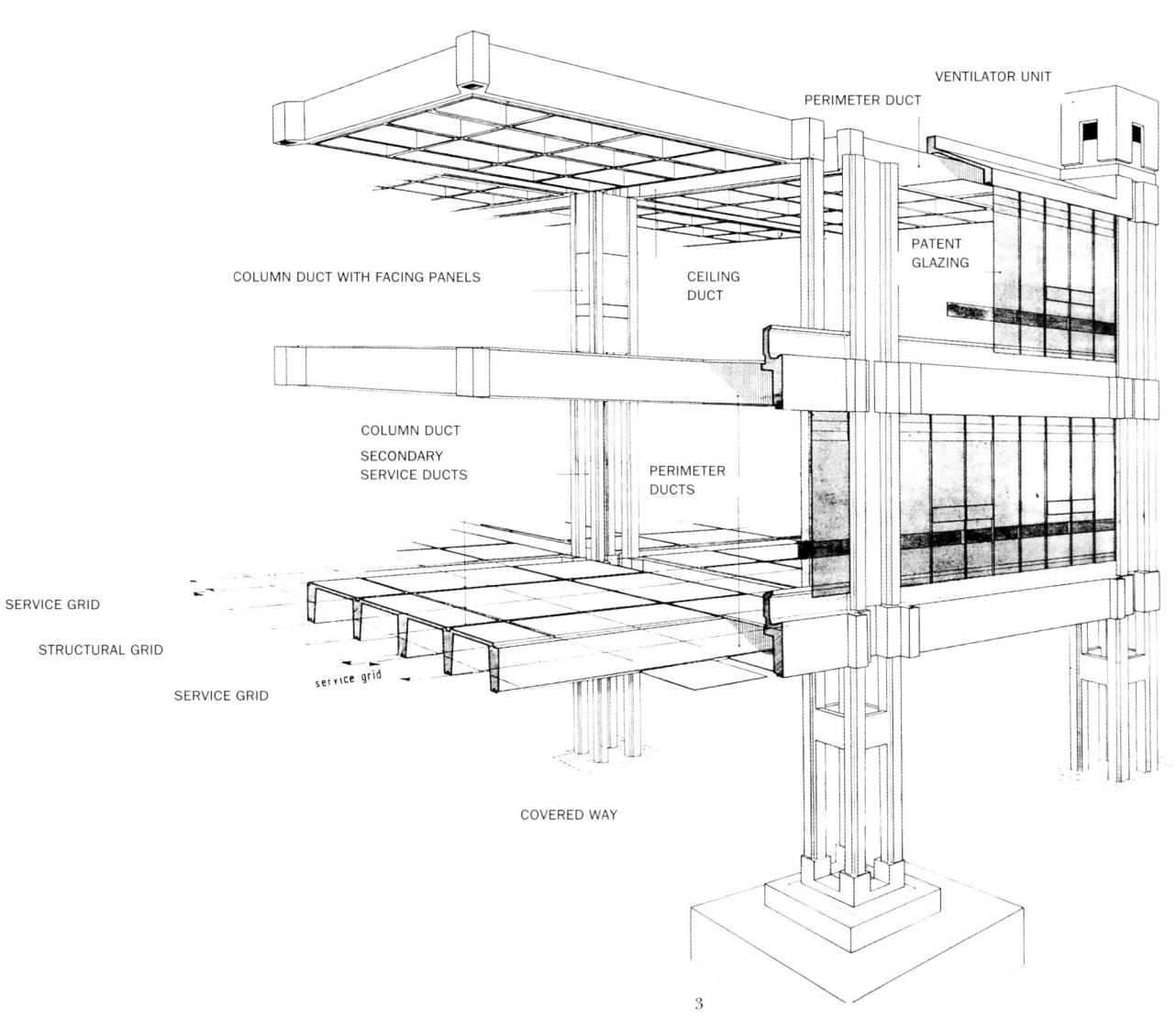

of servicing. Teaching rooms and laboratories are located on the perimeter for maximum daylight and natural ventilation, while circulation and dark rooms are at the core.

The elevations echo the structural pattern: habitable bays are glazed overall with glass, and service runs in concrete. Ventilation is by louvres, with pairs of opening lights in each bay to give a current of cool air. These accent the working areas in the same way as the concrete caps over vents and extraction fans emphasise the function of the service areas.

教室和实验室都布置在周边位置以获得最好的日照和通风，其他房间则布置在中心位置。

从剖面图可以看出其结构形式：开间适于居住，采用玻璃作为外墙和隔断，供应管线埋设在混凝土之中。设置散热窗以利通风，每个开间内还都布置数个冷气供应口。这些都改善了工作区的环境。通风装置和排气风扇的设置也增强了设备区的功能。

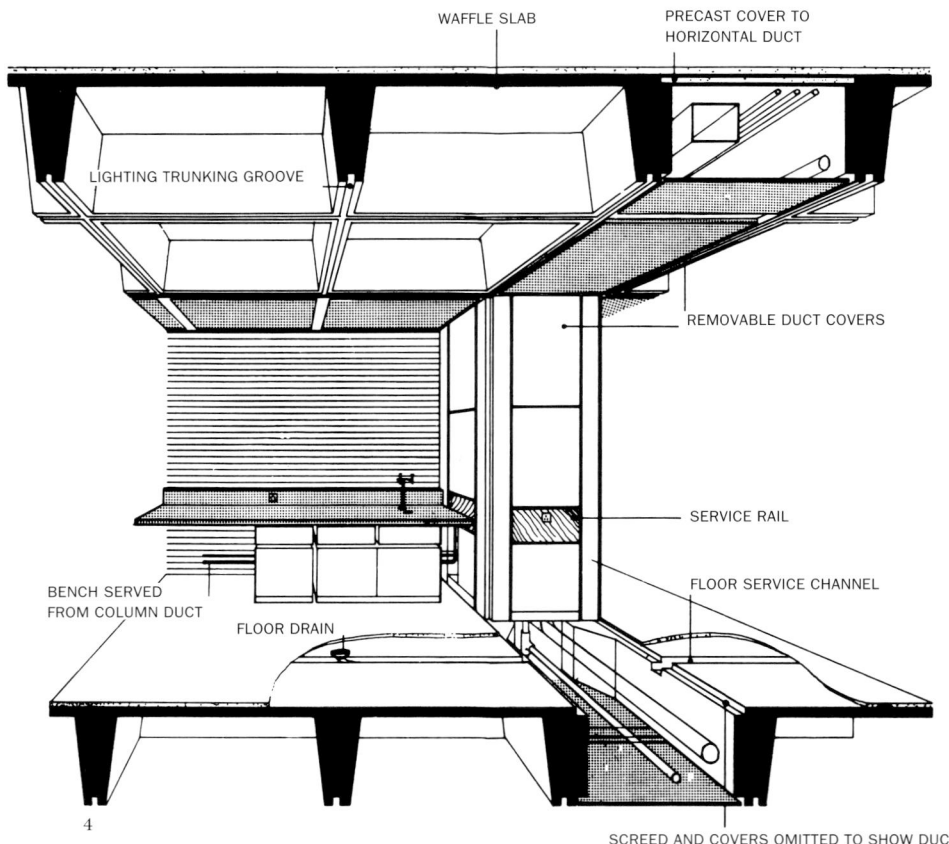

4

4 剖面分解图
5 结构和设备空间模型：混凝土板搁置在四根空心柱腿上，中空部分是为了保证设备管线连续的布置
6 吊装到位之前的预制混凝土楼板
7 楼梯按照正交方式布置
8 结构开间采用无油灰窗玻璃外装，设有玻璃散热窗

4 Exploded section
5 Model showing structure and service voids: a concrete table is placed between four hollow legs which provide continuous service runs
6 Precast concrete floor unit being lifted into position
7 Staircases are located by the discipline of the tartan grid
8 The structural bays are clad in patent glazing with glass louvres

5

6

7

8

Mining and Metallurgy, University of Birmingham 25

The Maltings Concert Hall

Design/Completion 1965/1967
Snape, Aldeburgh, Suffolk
Aldeburgh Festival
2,300 square metres
Brick
Roof: high tensile steel and wood, blue-black asbestos slates
Concrete floor, grit-blasted soft red brick, hardwood, sheradized steelwork

莫尔丁斯音乐厅

设计/完成　　1965/1967
斯奈普，阿尔德伯夫，萨福克郡
阿尔德伯夫音乐节
2,300m²
砖
屋盖：高强钢筋和木材，蓝黑色石棉瓦
混凝土楼板，喷砂处理的无矿盐红砖，硬木，镀锌钢材

In 1965 Benjamin Britten decided to explore the possibility of converting the largest malt house at Snape into a concert hall for the Aldeburgh Festival.

A brief was formulated for a concert hall to seat between 700 and 800 people, with lighting facilities for opera and an orchestra pit. In addition to festival performances, it was to be a recording studio during the remainder of the year, requiring a flat floor within the auditorium and facilities for stereophonic recording.

The plan proposed a single space accommodating the auditorium and full-width stage. It had already been decided to replace the existing roof and remove an existing screen wall to provide an auditorium width of about 20 metres. The roof design was central to the whole building. Externally it retained the shape and character of the original roof.

Continued

　　1965年，作曲家本杰明·布里顿计划将斯奈普最大的麦芽作坊改建成一个音乐厅，供阿尔德伯夫音乐节使用。

　　初步规划要求音乐厅的容量为700～800人，设有供歌剧演出之用的灯光设施和一个乐池。除了满足音乐节的需要，在以后的日子里它还可以作为录音棚使用，这就要求会堂采用平坦的地板，配备立体声录音设备。

　　规划中会堂和舞台包含在同一空间内，原有的屋顶和一面花格墙被拆除，改造成一座有20m宽的会堂。屋盖设计是整个项目的最核心部分，在外形上保留了原有屋顶的特征。

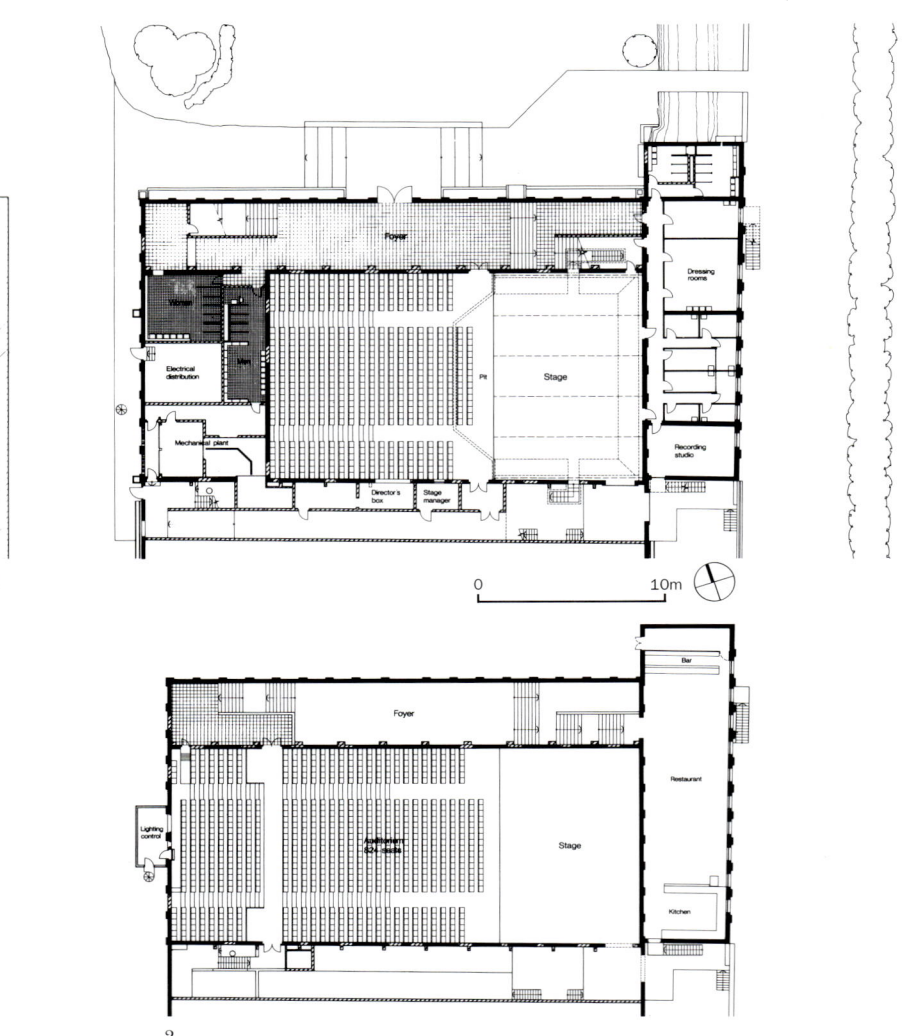

1　Entrance level plan
2　Upper level plan
3　The restoration of the maltings at Snape as a concert hall for Benjamin Britten and the Aldeburgh Festival introduced new uses and standards without damaging the character of the building
4　Cross section
5　The auditorium viewed from the stage

1　初步设计方案
2　修订后设计方案
3　根据本杰明·布里顿和阿尔德伯夫音乐节的需要，将斯奈普的一座麦芽作坊改建成音乐厅，建筑原貌依旧，但赋予全新的用途和功能
4　横截面
5　从舞台上看到的观众席

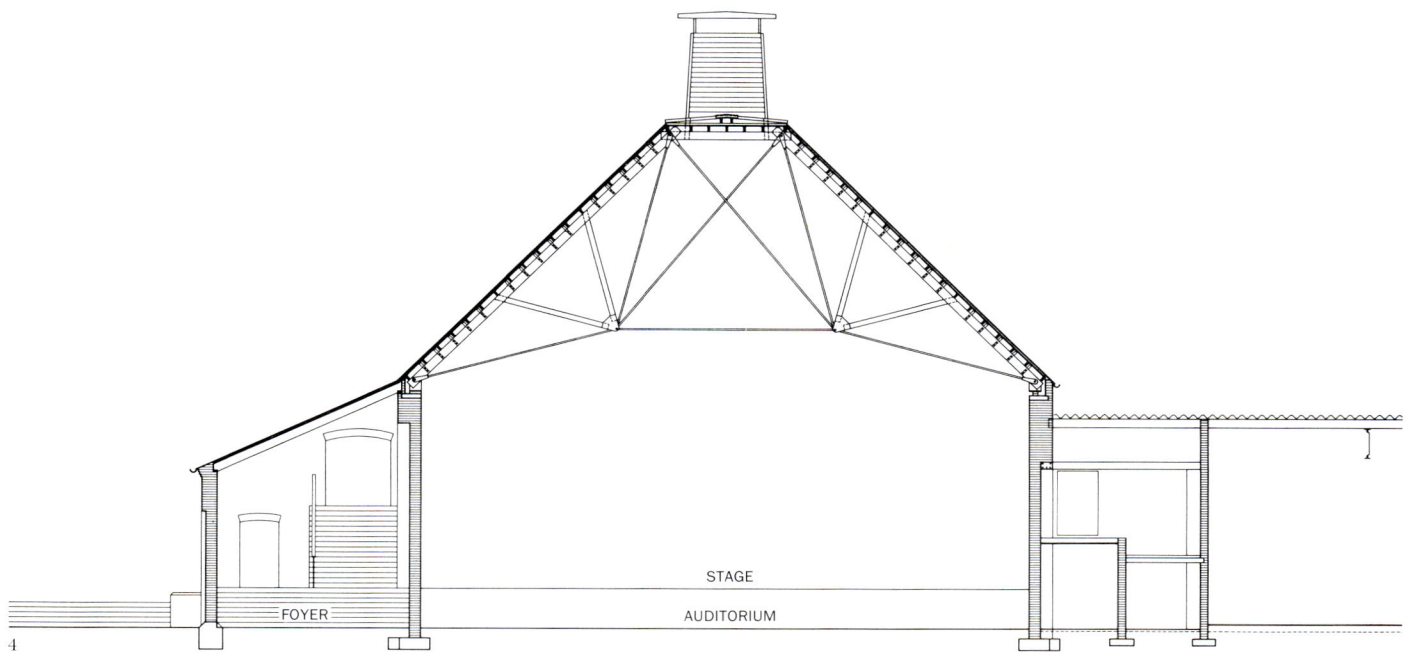

The Maltings Concert Hall　27

It consisted of a series of simple trusses at 4 metre centres of standard triangular construction. Compression members were in Douglas Fir and all the tension members in high tensile steel. This design allowed the cross-braced centre section to be adjustable to take up dimensional variations in the width of the hall.

To maintain the industrial character of the maltings, finishes were used sparingly. Brick walls, grit-blasted to reveal their original colour, provided increased sound absorption. Roof timbers were unpainted, the steel sherardized only. Seating, designed after Bayreuth, was in ash and cane.

The auditorium was opened on 2 June 1967. On the night of the first concert of the 1969 Festival it was completely destroyed by fire. Benjamin Britten decided that it must be rebuilt immediately, and within 42 weeks the concert hall was rebuilt to the original design.

6

屋盖为桁架式，由一系列标准的 4m 三角形拱架组成。压杆使用花旗松木，受拉杆件使用高强钢筋。这样设计的好处是，横截面中的三角形拱架可以进行调整，适应大厅不同段宽度变化的需要。

为了保持这个麦芽作坊原来带有的工厂特征，装饰设计非常谨慎保守。砖墙进行喷砂处理以展现其本来的颜色，同时也可以加强吸声效果。屋顶的木材杆件都不用油漆，钢材也仅仅镀锌处理。座椅仿照德国拜伊罗特歌剧院设计，使用岑木和藤条编制。

音乐厅于 1962 年 6 月 2 日开放，然而在 1969 年的音乐节的第一场音乐会上，它毁于一场大火。本杰明·布里顿决定立即重建这座音乐厅，因此 42 周之后，这个音乐厅按照原来的设计原样重建。

6　屋盖桁架的拉压杆件之间的连接
7　仿照拜伊罗特歌剧院设计的座椅，构成了室内空间的重要部分
8　音乐厅和录音棚还需要进一步建造，重修屋顶和配置更复杂的设备管线

6　Junction between compression and tension membranes in the roof truss
7　The chairs and seating, modelled on those at Bayreuth, form an important part of the interior
8　The concert hall and recording studio required extensive reconstruction, a new roof and the addition of sophisticated building services within an East Anglian malt house

8

IBM (UK) Limited Computer Centre and Assembly Plant

Design/Completion 1969/1974
Havant, Hampshire
IBM (UK) Limited
97,500 square metres
Steel frame, concrete
Production area: precast concrete panels with white flint aggregate and bronze tinted glass
Circulation spine: fully glazed steel frame

IBM（英国）计算机制造装配中心

设计/完成　　1969/1974
哈万特，汉普郡
IBM（英国）有限公司
97,500m²
钢框架，混凝土
生产区：预制混凝土板，白火石集料，古铜色玻璃
服务中枢：钢框架，全玻璃外装

In 1969 Arup Associates were commissioned to prepare a development plan to create a new manufacturing centre for IBM on a 15-hectare site near the sea at Havant in the south of England. The complex includes the Computer Centre, the Systems Assembly Building (for the assembly and testing of computer parts), and a Materials Distribution Centre (for their storage and despatch). The buildings are sited on either side of a service and circulation spine, which contains the main entrance, a cafeteria and the powerhouse.

The Systems Assembly Building and Materials Distribution Centre are single-storey steel-framed spaces, which can be adapted to varying needs. Spaces are subdivided into fire compartments by service cores running at right angles to the main spine. The main floor level is raised above the existing site to allow for a services undercroft. The Systems Assembly Building is air-conditioned.

1969年阿鲁普联合事务所承接IBM的委托，为其设计并建造一座新的制造中心，地点是在英格兰南部汉普郡的哈万特的海边，面积15hm²。这一复杂的项目包括一座计算机制造中心、一座系统装配中心（用于计算机部件的装配和测试），和一座材料配给中心（用于存储和配送材料）。这些建筑位于服务中枢的两侧，服务中枢包括总入口、自助餐厅和发电厂。

系统装配中心和材料配给中心为单层钢框架建筑，可以根据功能需要进行空间的调整。供应管线沿着中脊线的垂直方向布置，空间的分割也与供应管线的布置相一致。主要楼面的高程比场地平面要高，以设置地下室以安置设备管线。系统装配中心配备了空调系统。

1

1 The assembly building with the distribution block in the foreground
2 The large site, bounded by a major road and adjoining developments, fronts the sea
3 The single-storey steel-framed structures were designed to be adaptable for growth and change

1 系统装配中心，前景为材料配给中心
2 场地面向大海，面积很大，周围建有一条主要道路及一些毗连的建筑
3 单层钢框架结构，设计考虑了未来的更新和改造

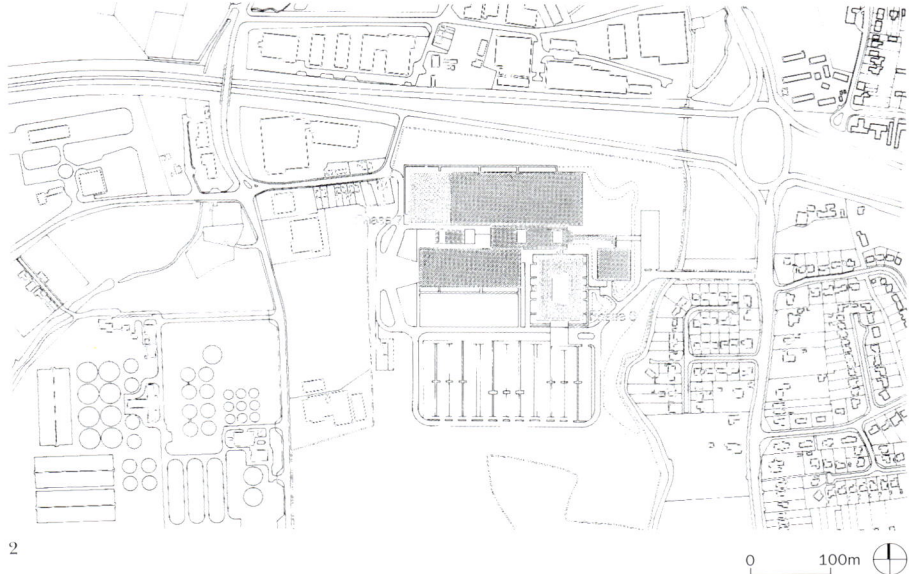

4 楼梯塔，采用玻璃外装
5 建筑群采用统一的围护系统，统一的基调中也有变化
6 从办公室里可以看到美观的中央庭院

4 Staircase tower and glazed link
5 The buildings are united by a cladding system that offers variations on a simple theme
6 The offices open onto a central landscaped courtyard

6 IBM (UK) Limited Computer Centre and Assembly Plant

7

7 A glazed central spine contains both the main entrance and shared central activities such as dining rooms, arranged around landscaped courtyards
8 A specially designed precast cladding system was designed for the manufacturing buildings
9 The glazed courtyard is protected by light metal sunscreens

7 服务中枢包括总入口和公用设施例如餐厅等，采用玻璃外装，围绕中央庭院而建
8 生产区建筑采用特别设计的预制板作为围护系统
9 为了避免玻璃给庭院造成过度日照，采用轻质的金属遮光帘

8

Selected and Current Works

作品精选

1970s

Horizon Project, Factory for John Player & Sons

Design/Completion 1968/1971
Nottingham
Imperial Tobacco Group Limited
125,055 square metres
Reinforced concrete, steel
Floor: in situ coffered floor slab
Precast lightweight insulating concrete; precast concrete cladding panels with white ridged and vertically tooled facing; bronze tinted glass in bronze anodised aluminium frames

地平线项目，约翰·普莱耶和后裔的工厂

设计／完成　　1968/1971
诺丁汉
帝国烟草集团有限公司
125,055m²
钢筋混凝土，钢材
楼板：现场镶铺楼板
预制轻质保温混凝土；预制混凝土围护面板，白色波褶饰面；铜色玻璃，铝合金窗框

The brief for this new factory demanded a building which would not only meet the client's present needs, but could readily accommodate change and allow the introduction of new manufacturing techniques. Essential requirements included large, column-free spaces with very heavy floor loadings, exactly controlled temperature and humidity, and high degrees of cleanliness and security.

The solution is based on a four-layered structural unit 30 metres square, with major columns at each corner. The ground floor is largely given over to car parking and despatch, while the first floor forms a continuous service void. These two layers have intermediate columns at 7.5 metre centres to support the loads from the machinery on the third floor—a clear space 6.5 metres high.

Continued

初步规划要求这一厂房建筑不仅应满足客户当前的需要，而且能容易迎合变化，适应不断采用的新生产工艺。要实现这样的目的，空间应该开阔且很少柱子的障碍（尽管荷载都很大），温度和湿度应该可以控制，清洁和安全应该能得到良好保障。

设计方案采用了一种正方的结构单元，这种单元边长40m，共4层，在四角设置主要承重柱。地面一层主要用于停车和货物配送，第二层为连续的供应服务区。这两层在中间7.5m处设置中柱，以承受第三层安装的机器设备的荷载。第三层高6.5m，不设中柱。

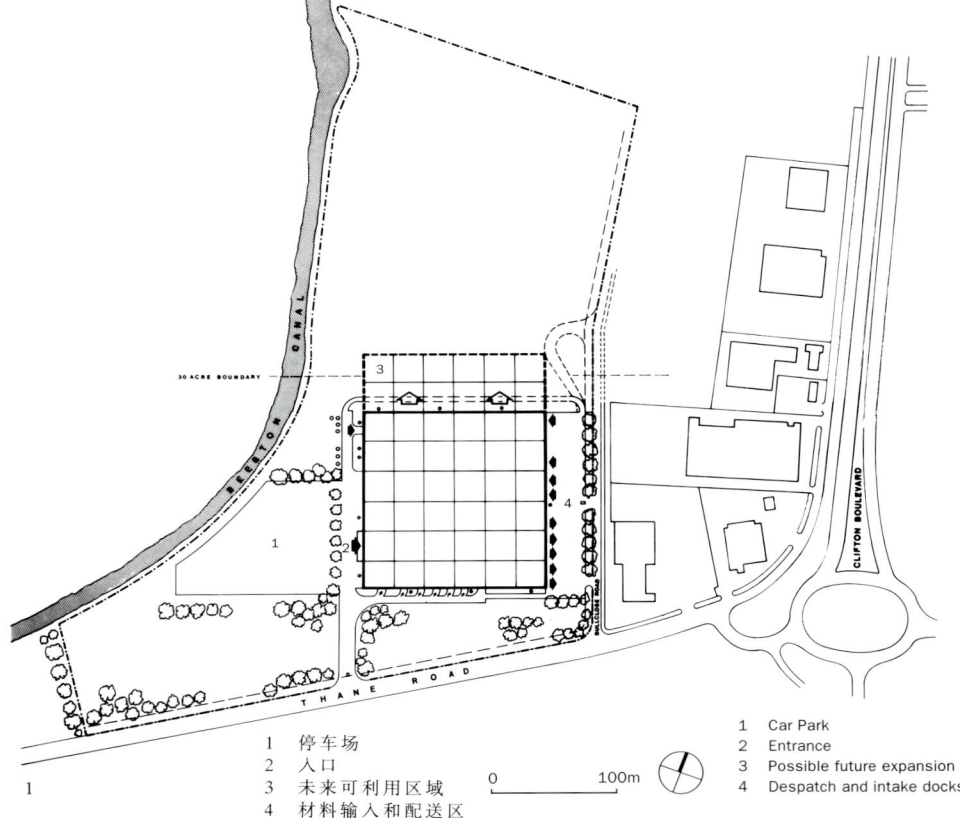

1　停车场
2　入口
3　未来可利用区域
4　材料输入和配送区

1　Car Park
2　Entrance
3　Possible future expansion
4　Despatch and intake docks

2

1　Site plan
2　Exterior facade
3　The organisation of the building is clearly expressed in the order of the elevation

1　场地平面图
2　外立面图
3　建筑的层次清晰地体现在正视图中

3
Horizon Project, Factory for John Player & Sons

All the air-conditioning plant and its associated ductwork is housed in the 5.7 metre deep roof structure. The building is made up of 36 units and can be extended simply by adding further units.

Power for the machinery, lighting and air-conditioning is generated on site by a total energy plant using gas turbine/alternator sets with North Sea gas as the prime fuel. Waste heat from the turbine exhausts is fed through boilers which provide steam for both manufacturing and air-conditioning processes.

The building also houses a range of facilities for staff including shops, a restaurant and a medical centre.

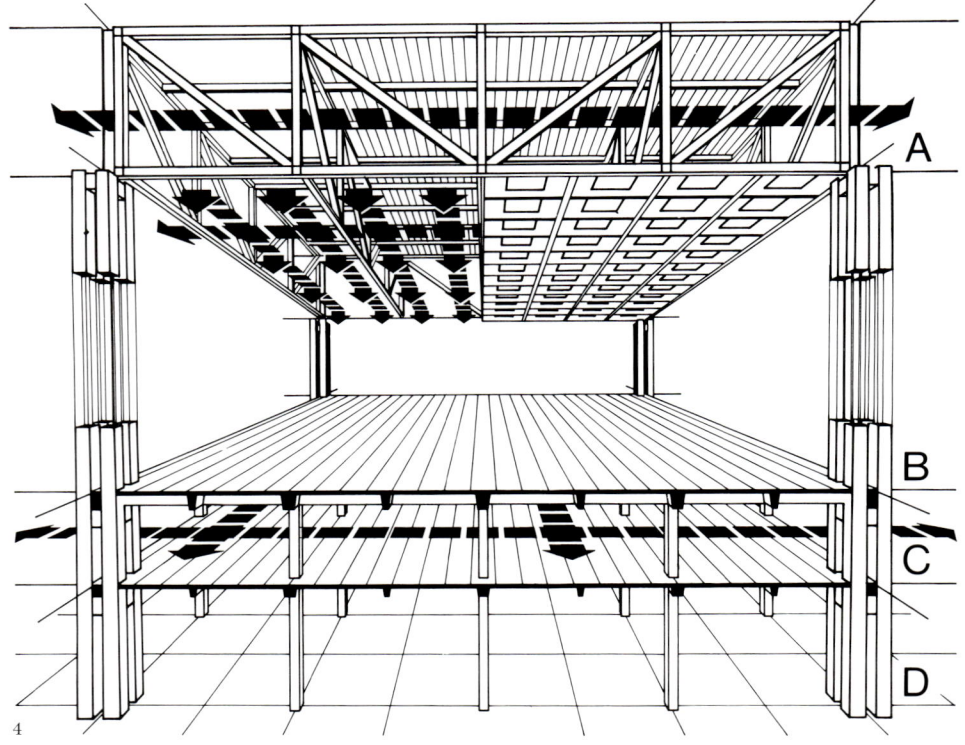

4

第四层是5.7m高的屋盖结构，所有空调设备和管道系统都安置其中。整个建筑由36个单元组成，如果要扩建的话，只要简单地增加更多的单元就可以。

机器设备、照明和空调系统使用的电力都由建筑附属的总发电厂供给。电厂使用涡轮交流发电机，采用来自北海的天然气作为燃料。涡轮发电机排出的废热通过锅炉制造蒸汽，为制造车间和空调系统直接提供动力。

建筑中还设置了一系列为工人和职员提供便利的服务设施，如商店、一家餐馆和一间医疗中心等。

5

6

4　The basic structural module is a 30 metre square unit within which the space is organised in four layers. This module is clearly expressed externally. The top layer (A) provides space for structure and servicing systems above the main production floor (B) with an interstitial service floor (C) and parking, vehicular services and amenities at ground floor level (D)

5　Production areas are large, flexible spaces lit by high-level glazing

6　The project was one of the first to be designed around a total energy plant

7　Study model

8　Staircases and services are kept outside the building envelope

9　The structural bay articulates the facade

4　结构的基本模数为边长30m的正方单元，空间分为4层。图中清楚表示了其外观，最顶层（A）安放结构和设备系统；（B）层为生产层，空间最为宽阔；（C）层为服务层，（D）层为停车场等便利设施

5　生产层空间十分宽敞，可以任意分隔；设有高窗提供良好的采光

6　建筑领先时代，配有附属发电厂

7　供研究用的结构模型

8　楼梯和设备管线设在围护结构之外

9　结构开间在正面接合

Horizon Project, Factory for John Player & Sons 41

IBM North Harbour

Design/Completion Phases I–IV 1970/1982
North Harbour, Portsmouth, Hampshire
IBM (UK) Limited
33,400 square metres
Steel and concrete, precast and in situ
Stepped landscaped terraces; precast concrete with exposed Portland Capstone aggregate in white cement, glazing
Steel framed glazed street

IBM 公司北港工程

设计/完成（I–IV 四个阶段） 1970/1982
北港，汉普郡
IBM（英国）有限公司
125,055m²
钢材和混凝土，预制，现浇
阶梯形斜坡露台；预制混凝土，采用波特兰石集料和白水泥，玻璃窗
拱廊设有顶棚，采用钢框玻璃窗

The IBM headquarters in the United Kingdom, planned to accommodate more than 2,500 people, was constructed on a 50-hectare coastal site reclaimed from the sea.

The project provides a wide range of facilities including computer rooms, workshops, offices, restaurants and major building services installations, and was planned in four phases.

The first two phases consisted of offices; a large computer centre was constructed as the third phase of development.

The fourth phase provides 34,000 square metres of space in a series of four stepped office buildings linked by a glazed arcade.

Continued

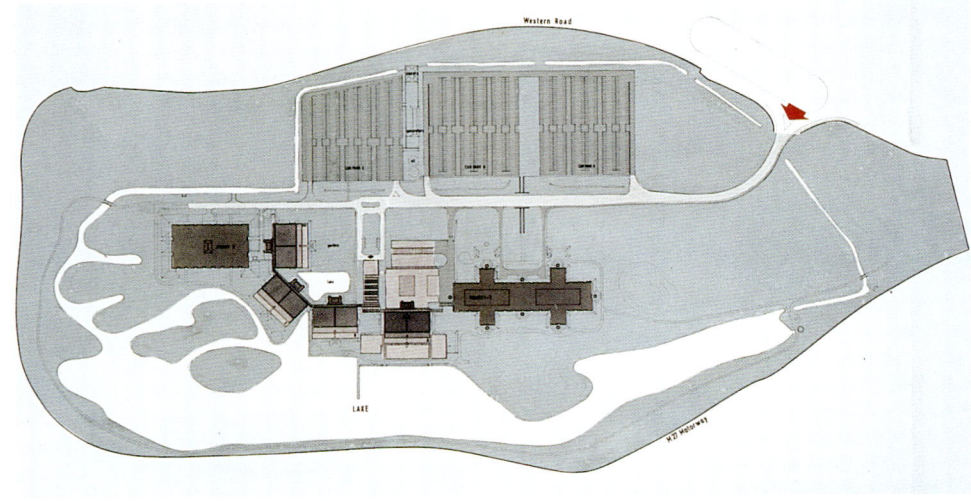

1

IBM 公司英国总部大楼计划容纳至少 2500 人，场地选在海边的一块填海面成的土地之上，面积约 50hm²。

建筑物内设施齐备，包括计算机房、工作室、办公室、餐馆和主办公楼。整个建造过程分四期进行。

前两期工程主要为办公室，第三期工程是建造一座大型的计算机中心。

工程的第四期是建造一组四级阶梯状的办公建筑，建筑面积约 34,000m²。建筑之间用拱廊相连，拱廊采用玻璃顶棚。

2

3

1 Site plan
2 The first two phases of development involved a four-storey office building built on the eastern end of the site
3 The main entrance hall
4 Cutaway drawing showing an office pavilion and glazed street in phase IV

1 场地平面图
2 前两期工程包括一幢四层高的办公建筑，位于场地的东端
3 主入口处门厅
4 局部剖面图，阶段 IV 所建造的办公建筑群和玻璃顶棚拱廊

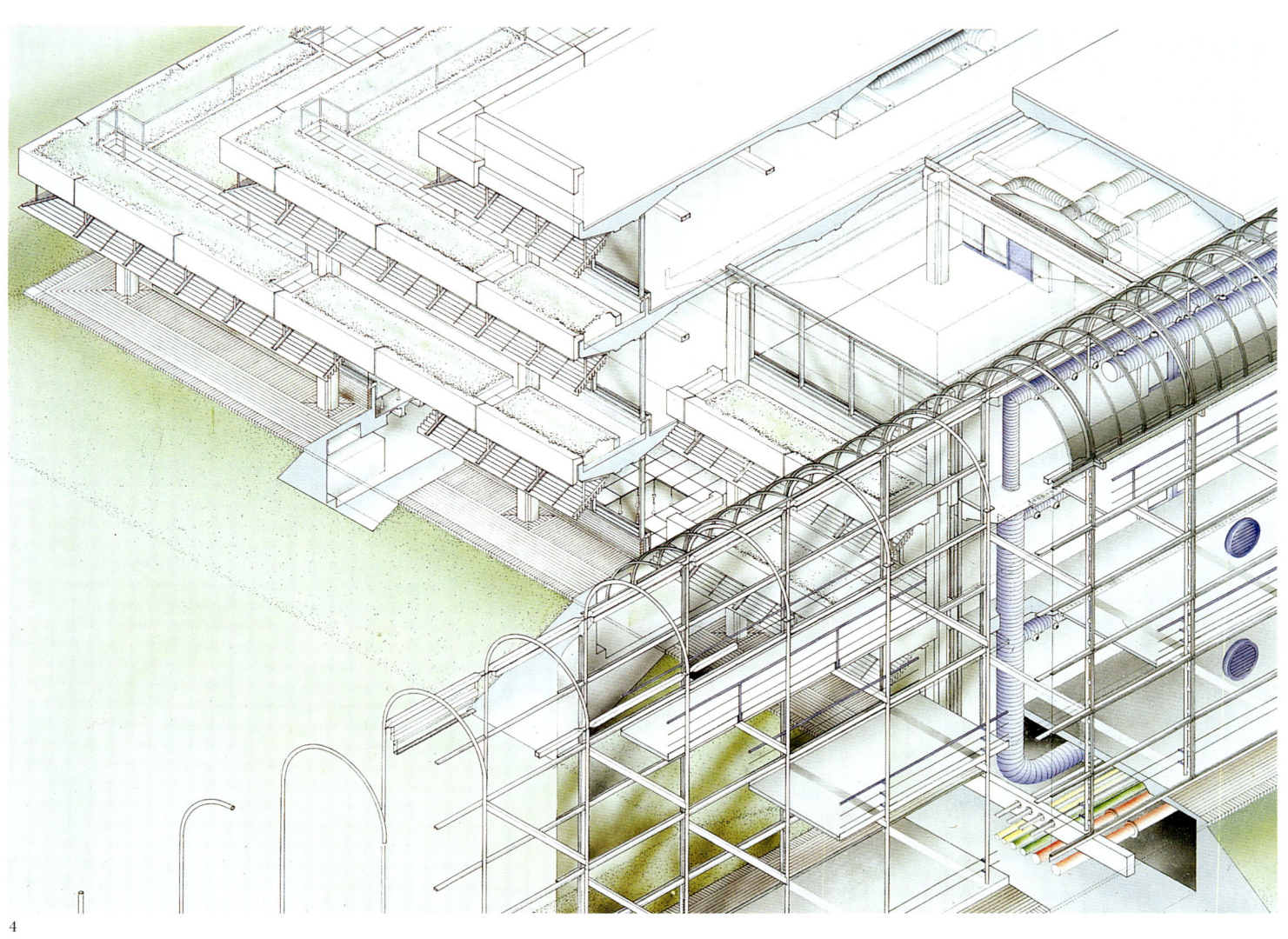

A new entrance, within a glazed pavilion, connects to the arcade and houses shops, exhibition areas and an auditorium.

Throughout the project the designers worked closely with the statutory authorities and IBM's own Real Estate Department. A strategy for large-scale landscaping was developed in conjunction with a consultant landscape architect.

Following recommendations by Arup Associates, major works of art were commissioned from Richard Smith, Robyn Denny and Howard Hodgkin.

入口处门厅采用玻璃顶棚和外装，从这里可以进入拱廊、一些房间、展示区和礼堂。

在整个项目的设计建造过程中，设计师们和管理机构以及 IBM 公司自己的房地产开发部门保持了密切的合作。建筑景观的设计方案是在与景观设计顾问公司的合作下完成的。

在阿鲁普联合事务所的建议下，主要的美术相关工作委托给了理查德·史密斯，罗宾·丹尼和霍华德·霍奇金。

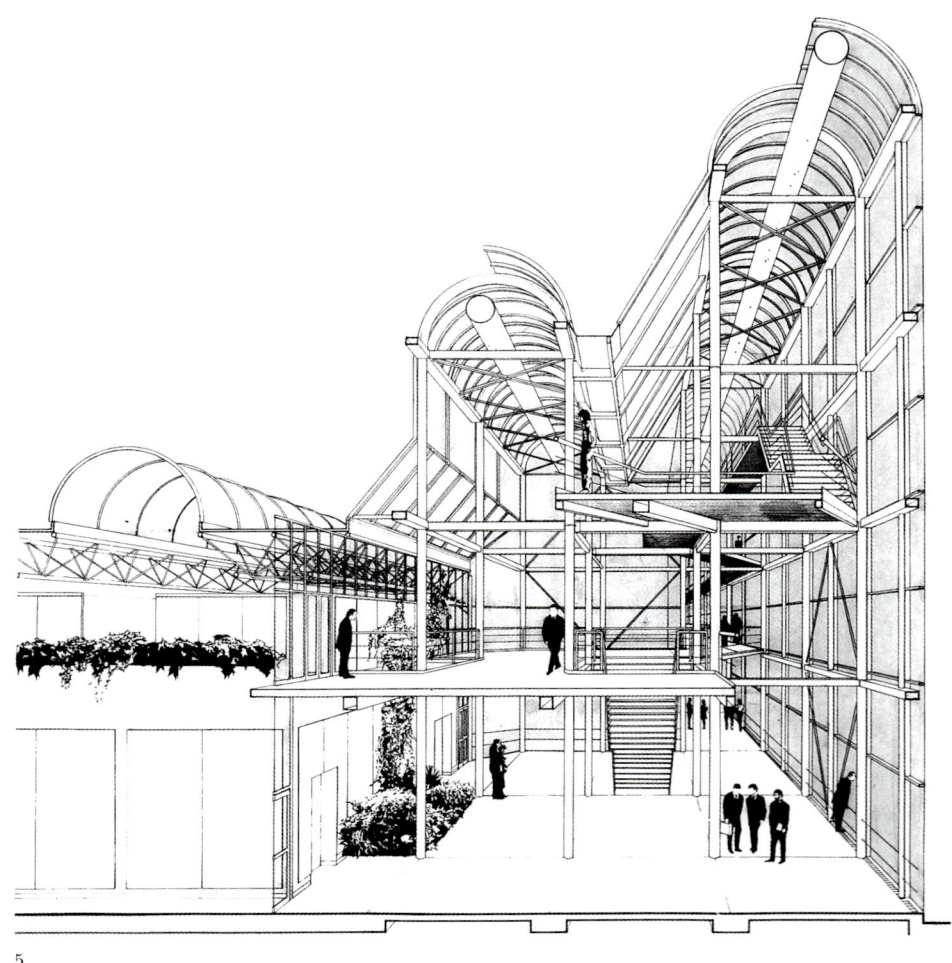
5

6

7

5 Section showing the entrance hall linked to the glazed street and stair hall
6 The main entrance hall is located within a glazed pavilion
7 Street interior

5 从断面中可以看到入口门厅与玻璃顶棚拱廊和楼梯间相连
6 主入口门厅采用玻璃顶棚和外装
7 拱廊内部

8

9

8　The reception area
9　Ramps link staff dining areas to coffee lounges overlooking the lake
10　The glazed street links four office pavilions

8　接待区
9　连接职员餐厅和咖啡室之间的坡道，从这里可以看到楼外的湖面
10　拱廊采用玻璃顶棚，将四级阶梯状的办公楼连为一体

Henry Wood Hall

Design/Completion 1972/1975
Southwark, London SE1
Southwark Rehearsal Hall Limited
1,300 square metres
Brickwork and quarry tiled floor
Lime and fibrous plaster interiors

In 1969 Arup Associates were invited to help find a suitable building for conversion into a rehearsal hall and recording studio for some of London's large orchestras. Holy Trinity Church in South London, designed by Francis Bedford and built in 1824, had been redundant since 1960, and was the chosen site.

A large, flat rehearsal space was created within the rectangular nave of the church and two recording rooms constructed below the gallery. The traditional lime and fibrous plaster was carefully reinstated to help retain the fine acoustics. The windows were double-glazed and the doors provided with acoustic seals. A perimeter ducted air supply system was installed by raising the nave floor level, with extraction through the ceiling light fittings. The large, brick-vaulted crypt was converted into a cafeteria.

Fine cast-iron railings were reinstated around the church and the original lampstandards were restored. New clocks, based on the original designs, were installed in the tower.

亨利·伍德音乐厅

设计/完成　1972/1975
南瓦克，伦敦东南1区
南瓦克音乐厅有限公司
1,300m²
砌砖，地板铺石砖
内部用石灰和加纤维灰泥抹面

1969年，阿鲁普联合事务所被委托为伦敦的一些大型交响乐队物色一座合适的建筑，以改造成一座演出用音乐厅和录音棚。最后阿鲁普联合事务所选中了位于伦敦南部的圣三一教堂（Holy Trinity Church）。这座教堂建于1824年，由弗朗西斯·贝德福德设计，随着时代的变迁，自1960年后它的用处就已不大了。

教堂中部原来的长方形听众席区域被改造成宽敞平坦的演出区，在教堂楼座的下方建造了两个录音棚。建造者用传统的石灰和加纤维灰泥材料小心的复原了内部墙面，以营造良好的声学效果。窗户采用双层玻璃窗，门采用隔声构造。大厅内地面进行了架高，以便安装沿建筑的四周布置的通风管道，顶棚的照明系统同时也进行了改造。砖砌拱顶的教堂大地下室改建成了一间咖啡室。

在教堂的周围装配了漂亮的铸铁围栏，原来的灯柱进行了更换。新的大钟仿照原来的设计制作，安装在塔楼上。

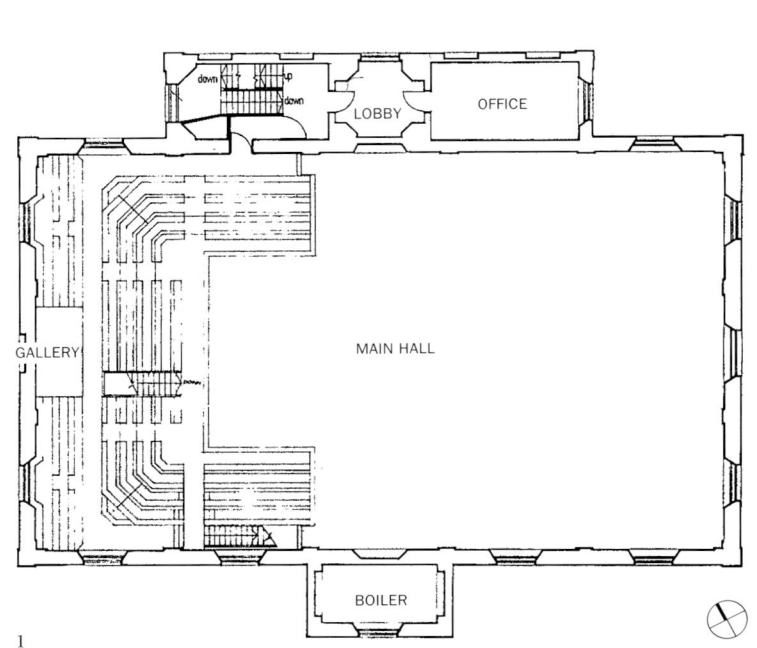

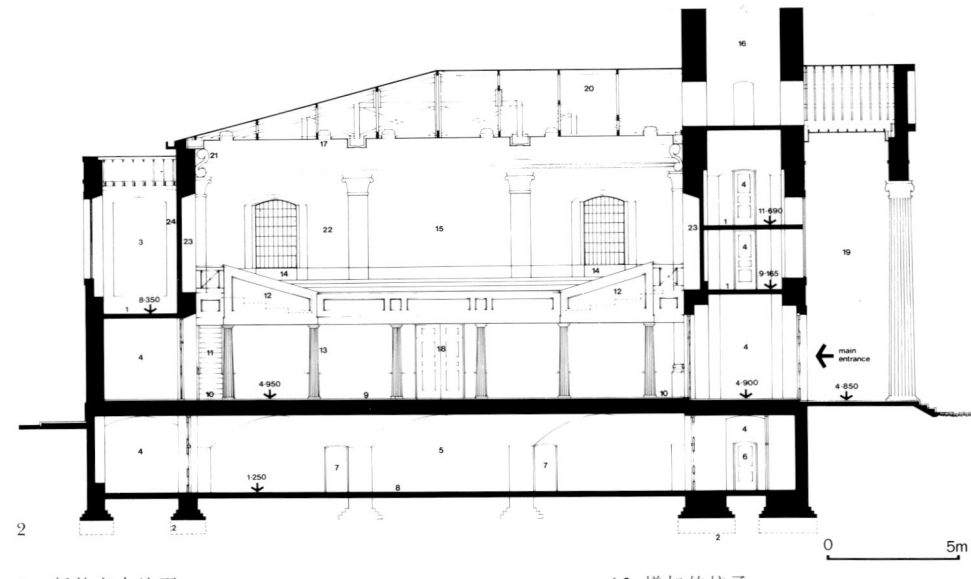

1 新的室内地面
2 加固后的建筑基础
3 锅炉房和贮水库
4 门厅
5 咖啡室
6 管理室大门
7 藏书室大门
8 混凝土地板
9 木材和石块地板
10 地板内沿四周埋设的通风管道
11 通往楼座的新楼梯
12 楼座的两边进行了改动
13 增加的柱子
14 双层玻璃窗
15 风琴陈设处
16 风通过塔楼通风
17 通过顶棚排风
18 录音棚大门
19 入口处柱廊
20 屋盖内空间
21 重新复原的中世纪式样顶棚
22 墙壁进行复原和粉刷
23 原来的窗户用砖砌封闭
24 墙内设置锅炉烟道，通往屋顶

1 Site plan
2 Section
3 Interior
4 The entrance portico and tower of the historic church
5 The brick-vaulted crypt was converted into a cafeteria
6 The work combines the conversion of the building for modern uses with the restoration of authentic detail

1 场地平面图
2 剖面图
3 内部
4 入口处的门廊和古教堂的塔楼
5 砖砌拱顶的地下室改造成了一间咖啡室
6 工程将基于现代使用的建筑改造和古建筑细节的真实复原融为一体

3

4

5

6

IBM Building, Johannesburg

Design/Completion 1969/1975
Johannesburg, South Africa
Soresons Property Limited
22,000 square metres
Reinforced concrete frame
Bronze glass, glazed brown tiles

This 22-storey office building is located on a site of half a city block in the centre of Johannesburg. Each of the floors is approximately 1,000 square metres; the majority are planned for use by IBM.

Above the main entrance and reception area are 22 storeys of offices. These floors are planned on a 1.52 metre module and can be laid out with a mix of open-plan and individual offices. Lifts and lavatories have been collected into a service tower on the north-east corner of the building. One floor is reserved for a computer installation and there are four levels of car parking below the entry level.

The building has two skins: the internal skin is an extension of the flexible partitioning system; the external skin of sun-absorbing glass is drawn over the structure, ducts and floor voids. The air-conditioning system allows for a maximum of 50 per cent glazing, as appropriate to the individual floor plans. Access and maintenance are provided within the ventilated cavity.

IBM 公司大楼，约翰内斯堡

设计／完成　1969/1975
约翰内斯堡，南非
索桑斯房地产有限公司
22,000 m²
钢筋混凝土框架
铜色玻璃，褐色上釉瓷砖

这幢22层的办公大楼位于南非约翰内斯堡市中心，占地约 1,000 m²，大部分供 IBM 公司使用。

主入口和接待区上方的22层都是办公区，这些楼层设计采用 1.52 m 的结构模数，可以按照开敞式平面布置办公室，也可以分隔出私人办公室。电梯间和厕所都集中在建筑东北角的设备塔楼处。有一层专门用作安放计算机设备，在入口处地下还有四层用作停车场。

建筑物设有两层围护系统，内层是开间隔断及其延伸出的外墙，外层是将结构、管道和构造完全包裹的玻璃幕墙，幕墙采用可以吸收阳光的玻璃。大楼配备的空调系统最大允许50%的开窗面积，这对于私人办公室来说是合适的。建筑也设有良好的通风管道系统。

1

1 地面一层接待区，通向楼外的街道
2 瓷砖饰面的入口顶棚，电梯和楼梯塔楼在玻璃幕墙的反射下闪闪发光
3 办公塔楼外围护采用双层的玻璃，电梯和楼梯塔楼外围护采用瓷砖饰面
4 地面一层陈列室，与入口相接

1 Reception at street level
2 The tiled entrance canopy and lift and staircase tower are glistening forms against the glass enclosure
3 The double-glazed skin to the office tower with tiled lift and staircase core
4 Ground-floor showroom adjoining the entrance

IBM Building, Johannesburg 51

Bush Lane House

Design/Completion 1970/1976
Cannon Street, London EC4
Trafalgar House Developments Limited
6,400 square metres
Concrete and steel
Tubular stainless steel and grey glass, anodised dark grey aluminium

布什·莱恩大楼

设计/完成　1970/1976
坎农大街，伦敦东部中央4区
卡法格（Trafalgar）房屋发展有限公司
6,400m²
混凝土和钢材
不锈钢钢管，灰玻璃，电镀灰铝

Bush Lane House is a lettable office building in the City of London. The building was designed so that the London Transport Jubilee Line tunnels (planned to pass directly beneath it) could be constructed at a later date. This constraint, which severely restricted positions for foundations, generated a design solution involving a table base supporting an office building with an external lattice structure.

The building consists of eight floors of uninterrupted office space over a first floor plant room. The offices were designed to be adaptable to the layouts of different tenants. Air-conditioning is provided through the square ceiling panels, each of which also contains a light fitting. The external wall is designed to exclude traffic noise. The outer skin of glazing, together with the limited window size, ensures that comfortable conditions are achieved economically.

The external tubular stainless steel lattice structure is water-filled to provide the required fire resistance.

　　布什·莱恩大楼是一座位于伦敦市中心的办公建筑。在未来的规划中，这幢大楼的地下要挖设地铁隧道，从而对设计提出了特殊的要求，基础的位置受到了严格的限制。最终设计方案采用平板基础，支承上部的办公楼和外部的格构建筑。

　　建筑首层为设备房，以上8层为办公室。办公室可提供不间断空间，因此可以根据不同租户的需要调整平面布置。房间的顶棚镶板上安装了空调口和照明设备。外墙设置了隔断交通噪声的构造。外部安装玻璃幕墙，对外墙上玻璃窗的尺寸也进行了限制，这都能确保空调系统比较节能。

　　最外部设置的格构式不锈钢钢管结构是为了防火而设计的，钢管中都充满了水。

1　Perspective view of Bush Lane House
2　Bush Lane House occupies a prominent site in the heart of the City of London

1　大楼透视图
2　大楼位于伦敦市中心的显要位置

3–4 Detail drawings showing the external water-filled lattice structure, glazed skin and connections

3－4 细部图，表现外部的充水钢管结构，玻璃幕墙及节点

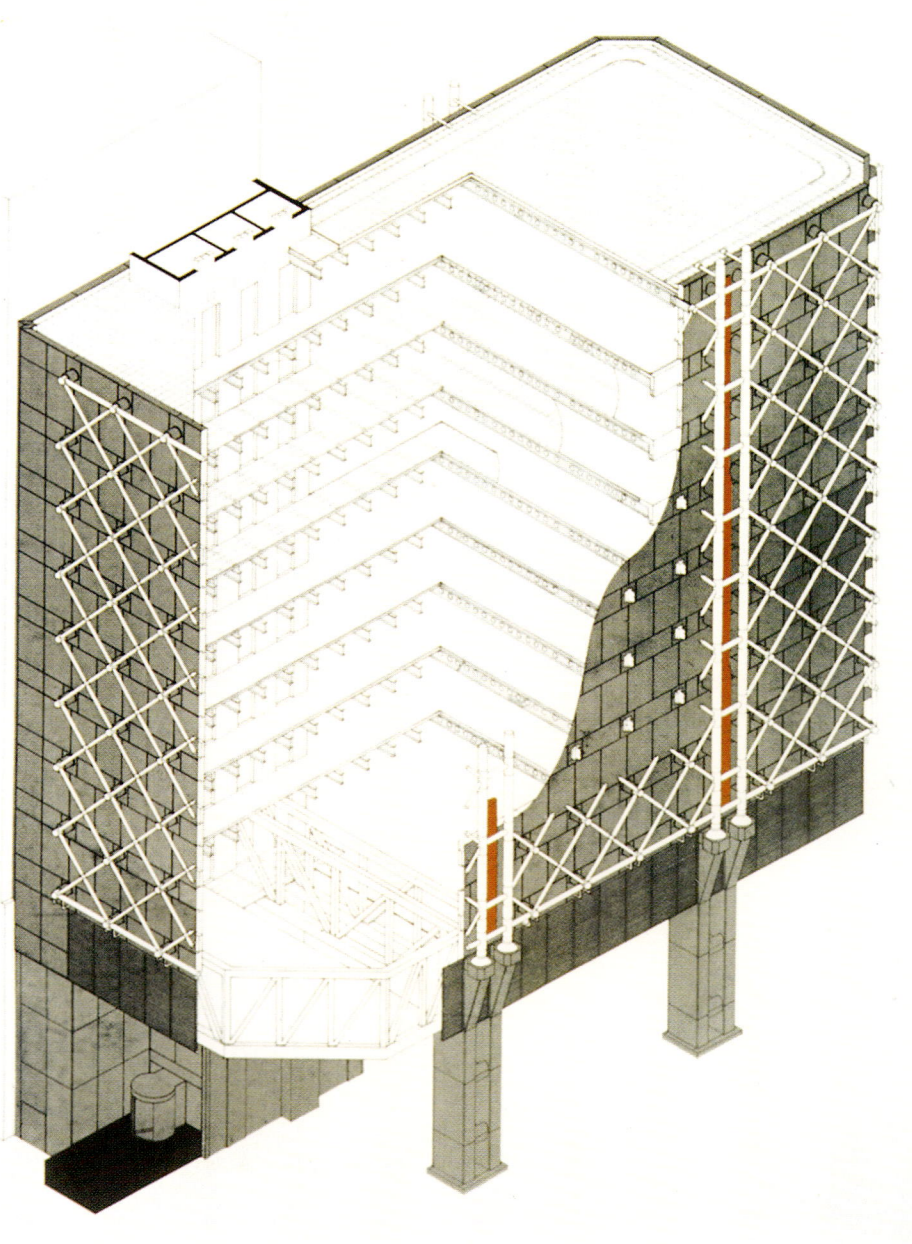

3

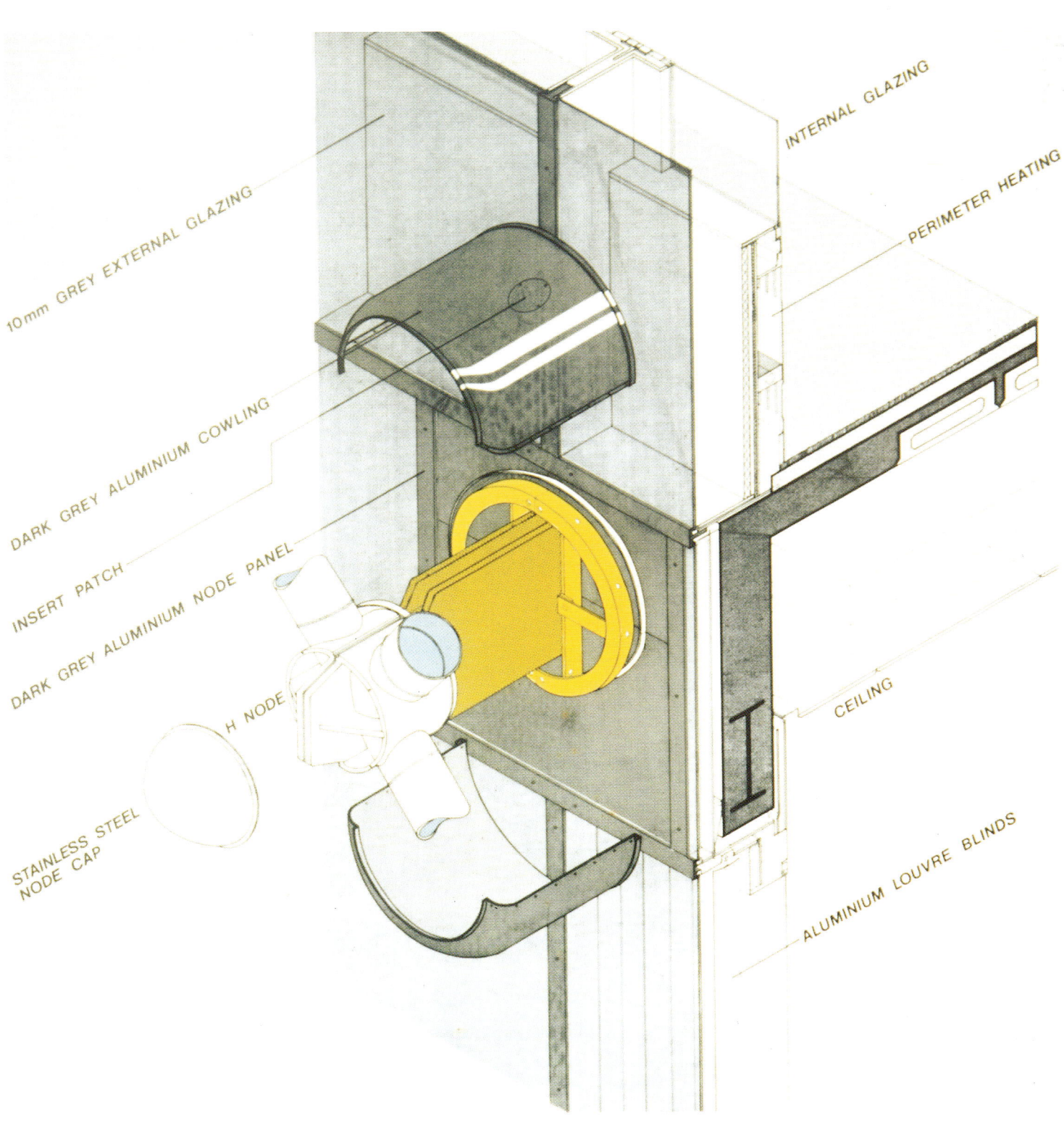

Gateway 1

Design/Completion 1973/1976
Basingstoke, Hampshire
Wiggins Teape (UK) PLC
19,000 square metres
Concrete and steel
Bronze anodised aluminium and bronze tinted double-skin panels

盖特威 1 号楼

设计/完成 1973/1976
贝辛斯托克，汉普郡
19,000m²
混凝土和钢材
铜铝合金和铜色双层板

The brief for a new air-conditioned head office required an efficient, low-cost-in-use building to house up to 1,000 people, offering flexibility of use and internal planning, with a variety of spaces suitable for open-plan and private offices.

The site is on a slope with unobstructed views to the south. This prompted a design which consists of five levels of office accommodation stepped up the slope, with terrace gardens at each level. Car parking space has been accommodated under the building at minimum cost by taking advantage of the natural contours.

In the interests of flexibility of use, speed of construction and economy, a repetitive 7.5 square metre structural unit was designed. This space accommodates the building services and is deep enough to allow these services to be installed and maintained without removing either the floor or ceiling panels, or disrupting offices above or below.

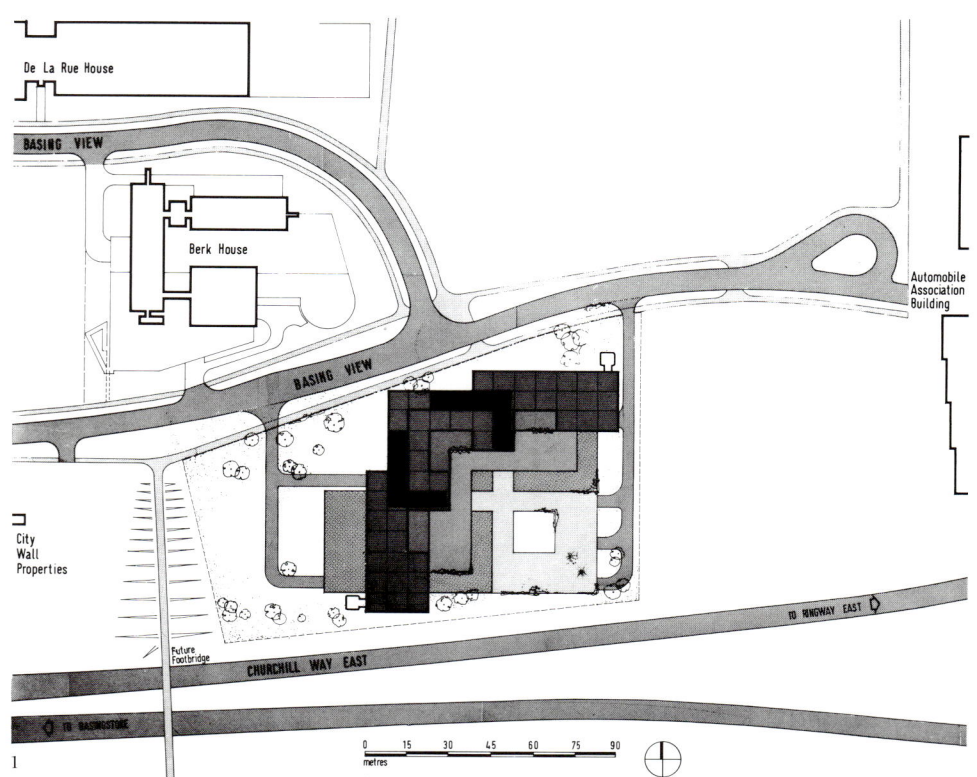

1

作为一幢公司总部大楼，设计要求建筑高效经济，容纳 1000 人，配备空调。内部空间在使用和布置上能有较大的自由度，可以根据不同需要分隔出开敞式办公室和私人办公室。

场地坐落在斜坡上，南面没有障碍而视野开阔。这启发设计师们提出了一个这样的方案：建筑沿斜坡分五级成梯形布置，在每一级的阶地上设计了花园。停车场设在建筑的地下，且利用了场地天然的高差条件，因此花费比较经济。

考虑到空间利用的自由度、施工周期和经济成本，建筑采用了重复的结构单元，这种正方形单元边长 7.5m。容纳设备管线的空间置于地下深处，使得设备管线的安装和维护都不用移动地板和顶棚，也不会扰动附近的办公室。

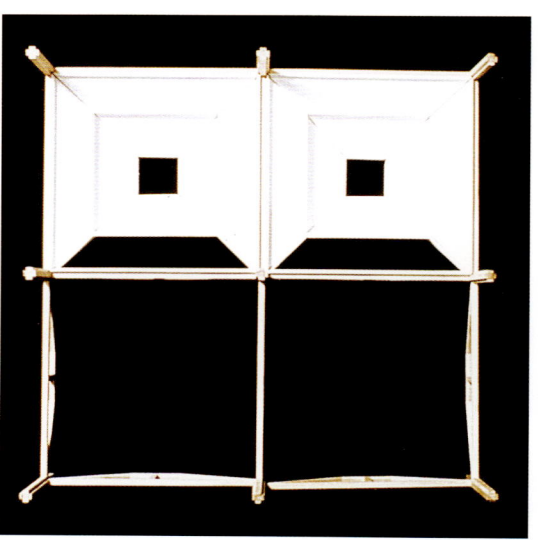

2

1 Site plan
2 The planning discipline derives from the structural bay which is 7.5 metres square. The pyramid form provides spatial definition in offices and creates spaces for services
3 Landscaped roofs create a series of stepped terraces and gardens

1 场地平面图
2 设计以7.5m正方单元为基本结构单元。层叠的建筑形式为办公室提供了三向的空间划分，也为设备管线提供了空间
3 每一层建筑的屋顶平台都进行了景观设计，营造出呈阶梯状的花园

3

4　The plant room is situated alongside the main entrance
5　Exterior terrace
6　Exterior
7　Sections A and B

4　设备房布置在主入口的旁边
5　花园阶地外景
6　外景
7　剖面 A 和 B

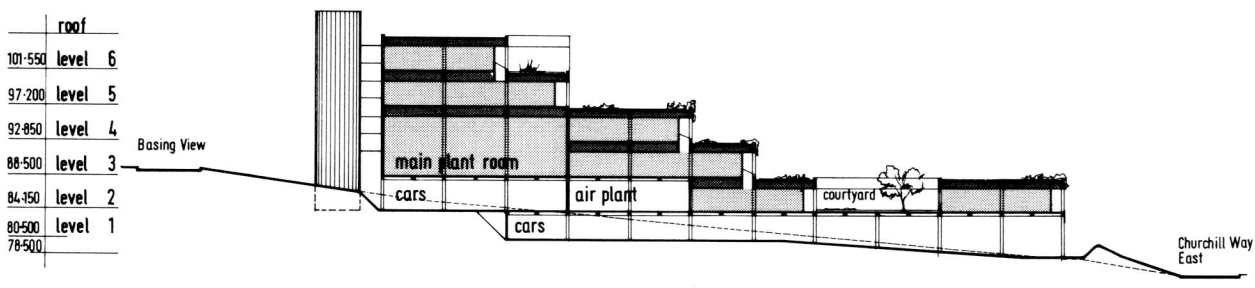

SECTION A

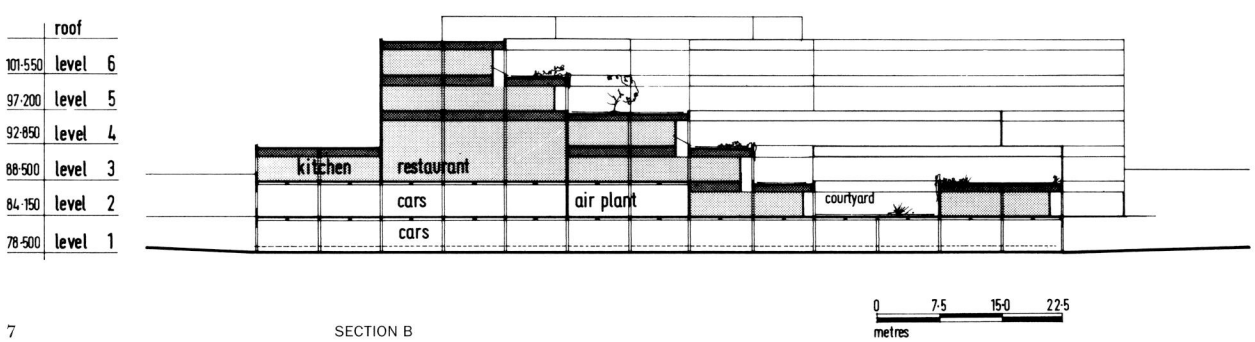

7　　　　　　　　　SECTION B

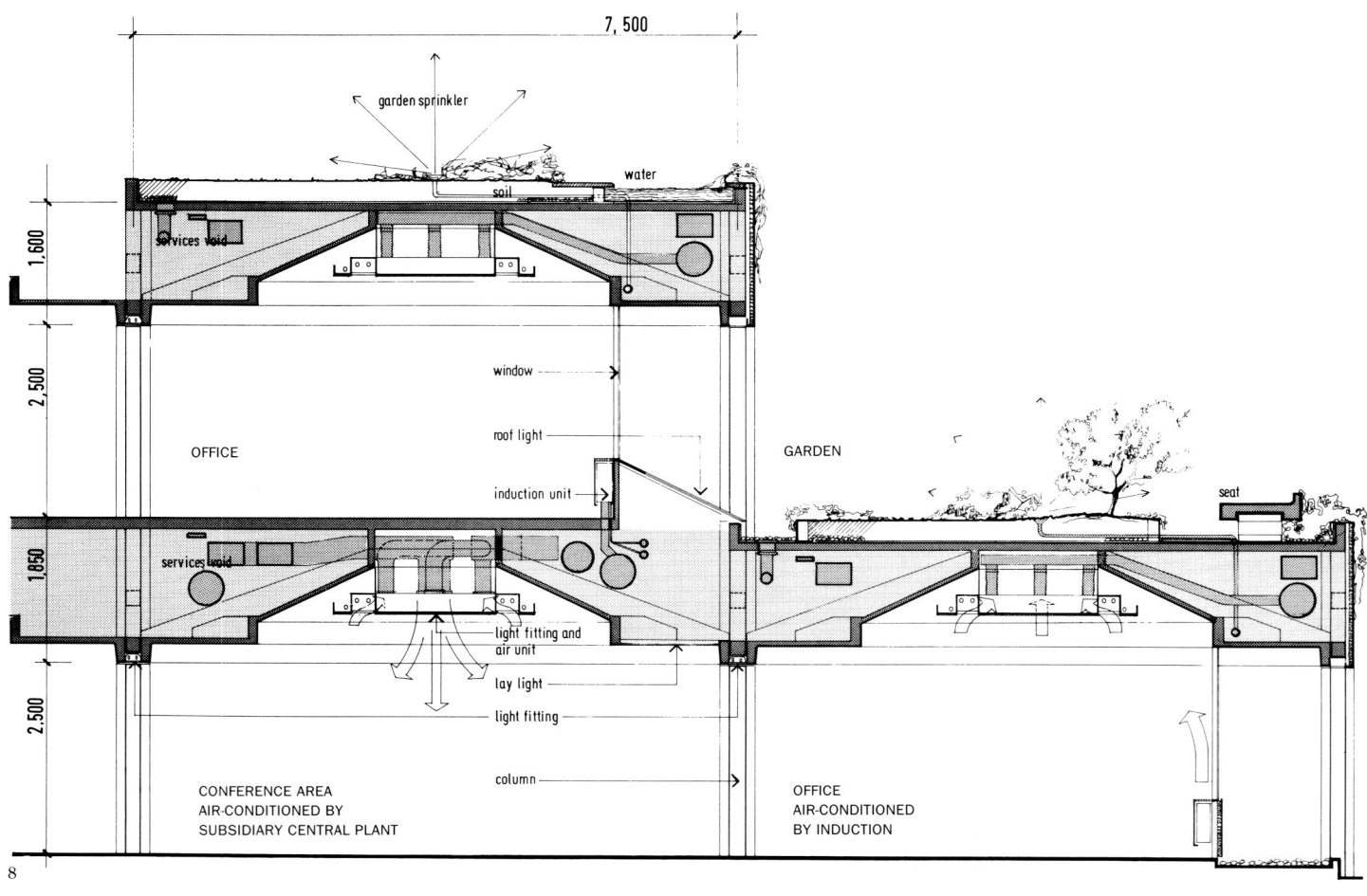

8　Section drawing
9　The stepped form of Gateway 1 provides an outlook onto gardens for people working in the building

8　剖面图
9　阶梯层叠形式使得工作者从楼内就可看到窗外的花园

Truman Limited Headquarters

Design/Completion 1972/1976
Brick Lane, London E1
Truman Limited
15,000 square metres
Brick, concrete grid ceiling, glass fibre reinforced cement floor
Brick and reflective glass

杜鲁门有限公司总部

设计/完成　1972/1976
砖巷，伦敦东1区
杜鲁门有限公司
15,000m²
砖，混凝土栅格顶棚，玻璃纤维加强水泥楼板
砖，反光玻璃

The new headquarters for Truman Limited were built on the site of their original brewery alongside a Conservation Area at Spitalfields in London. The site includes two listed Georgian houses that have been retained, restored and incorporated as an integral part of the scheme. Together with the new building they enclose a new courtyard which is defined on the fourth side by the historic facade of the old stable buildings in Brick Lane.

Industrial spaces—workshops, stores and truck docks—are located on the lower two floors. Amenities including restaurants, bars and a staff social club, are located on the third floor within a brick podium. The office space is located on the next three floors. The brick podium forms an extension to one of the historic buildings, following original street lines and maintaining the scale of its surroundings.

Continued

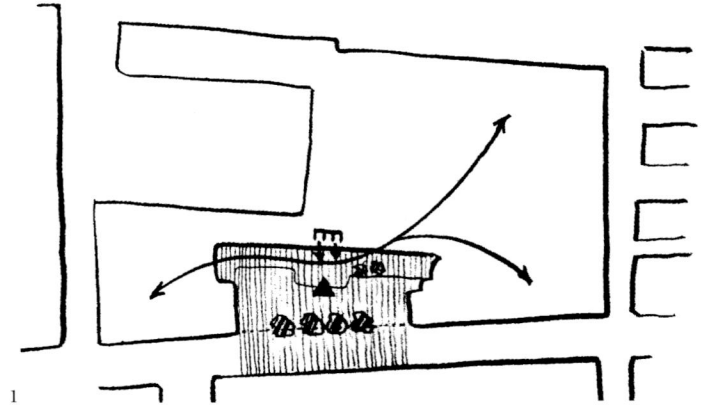

杜鲁门有限公司的新总部大楼建在伦敦斯普特费尔德的城市保护区内，场地原来是公司的酿酒厂。场地区域内有两座被列入保护名单的乔治王朝时代的古建筑，设计方案中也包括对它们进行保护和修复，使整个区域的景观成为一体。这两座建筑，以及砖巷（Brick Lane）的老马场建筑的正面，和新建筑共同围成一个新的庭院。

工业用空间如车间、仓库和卡车停车场都设置在地下两层。休闲娱乐设施如餐馆、酒吧和职员俱乐部都设置在第三层（及地面一层）。办公用空间设在其余的三层。地面一层采用了砖砌的列柱和墩墙，与周围历史建筑的构造保持协调。

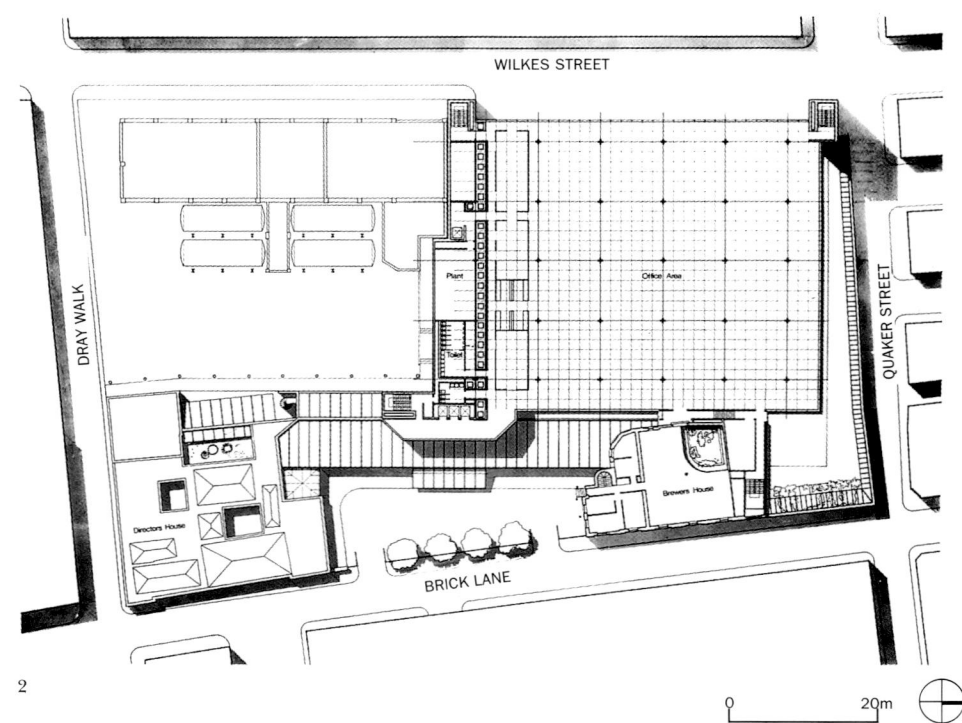

1 The plan was developed to create a new courtyard on Brick Lane—a focus of activity that linked old and new
2 Third-floor plan
3 Historic view of Brick Lane
4 Exterior view on Brick Lane looking west
5 The courtyard, framed by the restored Brewers House and the offices, preserves the scale of the historic setting

1 规划方案在砖巷设计建造一个新庭院，将新建筑与砖巷的历史建筑衔接起来
2 三层平面图
3 砖巷旧貌
4 由砖巷往西看到的情景
5 由复原后的酿酒厂建筑和新办公大楼围成的新庭院，维护了该区域原有的历史风貌

3

4

5

Truman Limited Headquarters 63

The office floors are stepped back from the street and are enclosed within a glass facade which cascades to the main entrance of the building. This forms a conservatory which links two Georgian houses and encloses a new paved courtyard.

The environmental services have been designed as an integral part of the structural system, with conditioned air distributed through structural concrete coffered ceilings.

办公区的楼层背靠街道呈阶梯状，正面外装采用反光玻璃幕墙，层叠状直至建筑的主入口，将入口门厅构造成了一个温室，连接起了原来的两座乔治王朝时代的古建筑，并围成了一个新的庭院。

设备管线与结构体系融为一体，空调口分布在混凝土顶棚之中。

6

6　发光的玻璃幕墙中可以看到周围的乔治王朝时代建筑
7　由庭院外部看到的情景

6　The mirrored glass facade reflects the surrounding Georgian buildings
7　Exterior view across the courtyard

8 Section showing stepped offices and entrance hall overlooking the new courtyard formed on Brick Lane
9 Section through the concrete coffered ceiling
10 An open-plan office floor showing the floor-to-ceiling glazing around the perimeter

8 断面图中可以看到层叠状的办公楼层、入口门厅，从上可以俯瞰依傍砖巷所建造的新庭院
9 混凝土顶棚断面，平顶镶板构造
10 开敞式平面布置的办公室，可以看到四周外墙采用的是高至顶棚的落地玻璃窗

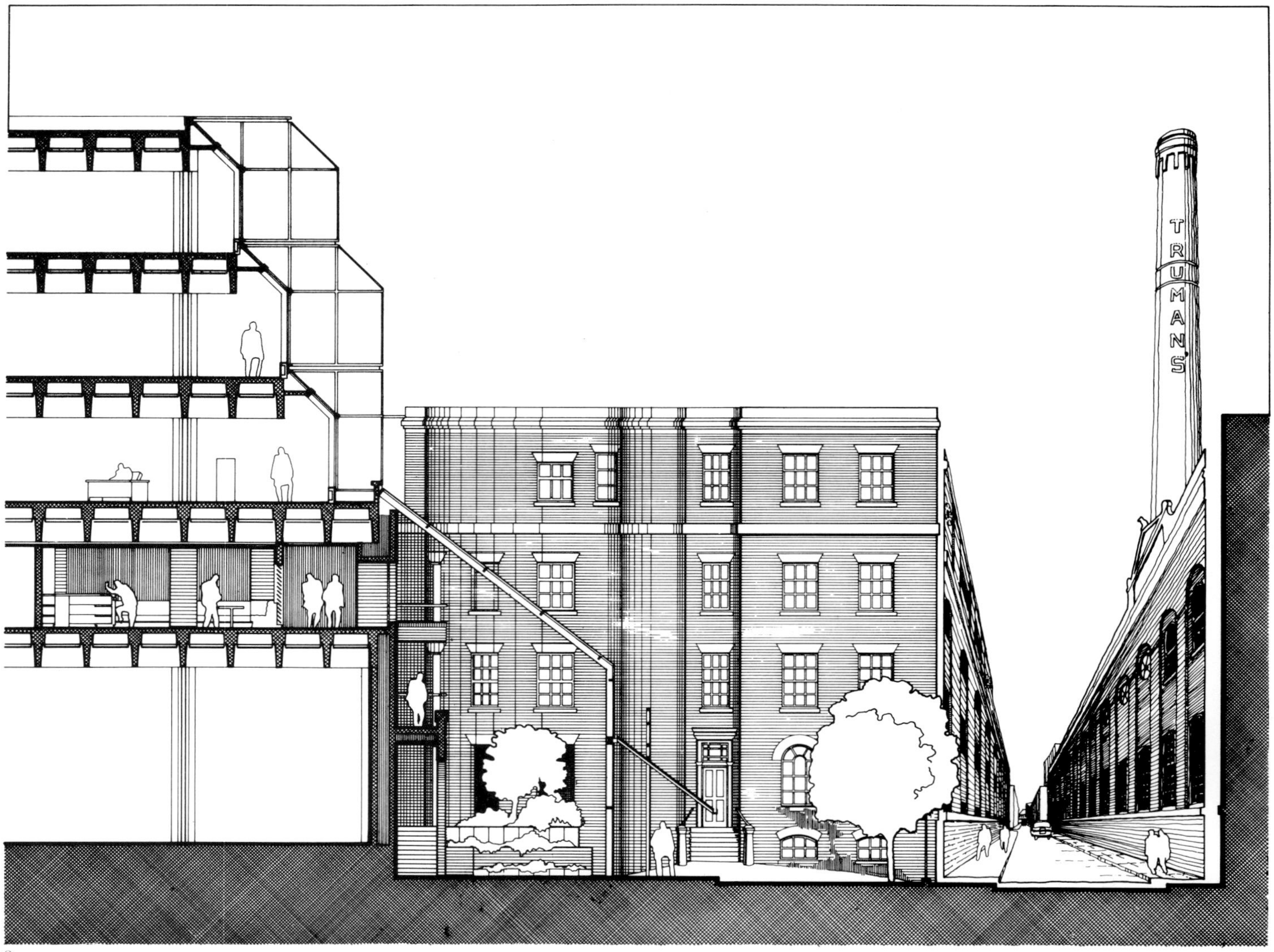

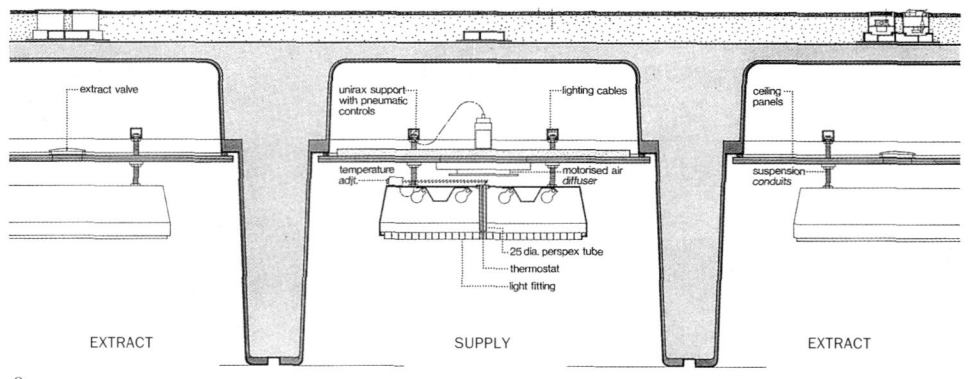

11

12

11 The perimeter route on one of the office floors
12 The entrance hall, an extension of the courtyard, forms a conservatory between the Georgian buildings
13 The bar interior

11 办公楼层四周的通道
12 入口门厅是庭院向建筑内的扩展,由玻璃幕墙构造成一个温室,和两侧的古建筑相连
13 酒吧内部

Sir Thomas White Building, St John's College, Oxford

Design/Completion 1970/1976
Oxford
St John's College
3,800 square metres
Precast concrete frames
Stone cladding, glass, aluminium frames

托马斯·怀特爵士大楼，牛津圣约翰学院

设计/完成　　1970/1976
圣约翰学院
3,800m²
预制混凝土框架
石材围护结构，玻璃，铝框

The Sir Thomas White Building at St John's College provides 156 rooms for graduates and undergraduates. Half are study/bedrooms and half are units containing a sitting room and a bedroom.

The site forms an extension to the college, adjacent to its very fine garden and existing stone buildings. The new building has multiple relationships: a street frontage to Museum Road; a garden frontage to the old garden; and an extension to the existing quadrangles. The entry to the building is visible from St Giles.

The building fulfils the college's current and future needs, but within a historic context. It is a modern building that reflects the mood of Oxford, the character of its surroundings and the silhouette of a medieval city.

Continued

1

位于圣约翰学院的托马斯·怀特爵士大楼为学院的本科生和研究生提供了156个房间。一半是学习室/卧室，另一半为包含一间起居室和一间卧室的单元楼。

大楼位于学院校园之内，毗连学院漂亮的庭园及原有的石建筑。新建筑体现了多重的与周边环境的关联，设计了与博物馆路相配合的街道，与原有庭园相接合的新庭院，有效地扩展了学院原有方庭的范围。从圣伊莱斯可以看到大楼的入口。

建筑很好地满足了学院当时及未来的需要，同时也承接了与历史的联系。这是一幢能反映牛津精神风貌，能与周围环境融合保持中世纪大学城风貌的现代建筑。

2

3

4

1　建筑规划模型，体现现有建筑之间的关系
2　大楼里可以俯瞰圣约翰学院的庭园
3　从圣阿尔丹斯看到的建筑外景
4　建筑外景局部

1　Model of development showing the context of existing buildings
2　The building overlooks the gardens of St John's
3　The view from St Aldates
4　Exterior view

Sir Thomas White Building, St John's College, Oxford　71

5

The form of the building has been developed as a series of framed, stepped pavilions. On each floor groups of four rooms alternate with solid bays enclosing stairs and services. The top floor is set back to provide balconies to the penthouse rooms.

A colonnade links the staircases and common rooms at ground floor level, and the circulation pattern emphasises the relationship of the new building to the existing college.

结构采用框架式阶梯形单元，每个单元即为一个房间。每个楼层每4个单元一组，间隔设置实体的开间，以设置楼梯及设备管线。建筑顶层单元内凹，以布置楼顶阁楼的阳台。

地面层设计了柱廊，将楼梯和一般的房间相连。柱廊的环形布置也是基于与学院原有环境相协调的考虑。

6

5　The clear articulation of the structure framing the undergraduate rooms
6　Athough a modern building, it reflects the medieval mood of Oxford
7　The aim was to create in each room what the President of St John's called 'a world within a world'
8　Floor plan
9　Exterior view of a cloister

5　结构框架单元的连接部位
6　虽然是现代建筑，同样也反映了牛津的中世纪精神
7　设计目标是使每个房间如圣约翰学院院长所说："世界之中的世界"
8　楼层平面图
9　回廊外景

7

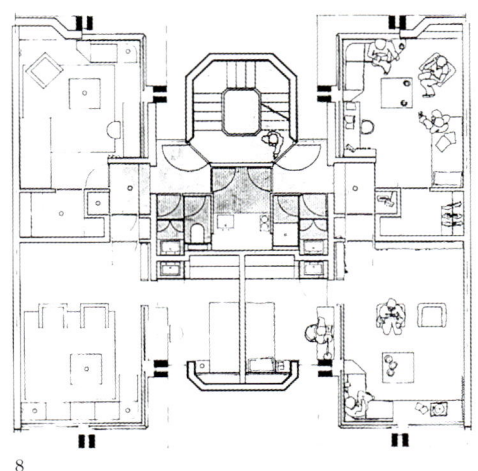

8

9

Sir Thomas White Building, St John's College, Oxford

Lloyd's of London Administrative Headquarters

Design/Completion 1973/1978
Gunwharf, Chatham, Kent
Corporation of Lloyd's
20,000 square metres
Precast concrete
Red brick, grey roof tiles, timber panelling

伦敦劳埃德保险公司总部大楼

设计/完成　1973/1978
冈沃夫，查莎姆，肯特郡
劳埃德保险公司
20,000m²
预制混凝土
红砖，灰色屋顶瓦片，木质镶板

In 1974 Lloyd's of London decided to move certain departments to a new building out of the City. The new site, once part of Chatham Naval Dockyard, was on the Medway with magnificent views towards Rochester Castle.

The brief was to design a building of about 20,000 square metres containing offices, data processing and ancillary activities, including a staff restaurant. It was also important to take full advantage of the characteristics of the site, particularly the river frontage and the views beyond.

The new building was planned around two landscaped courtyards. It is set into the slope of the site and in this way the apparent bulk of the building is reduced.

The four office floors were planned to allow changes in the size and location of departments. With the exception of the air-conditioned computer room,
Continued

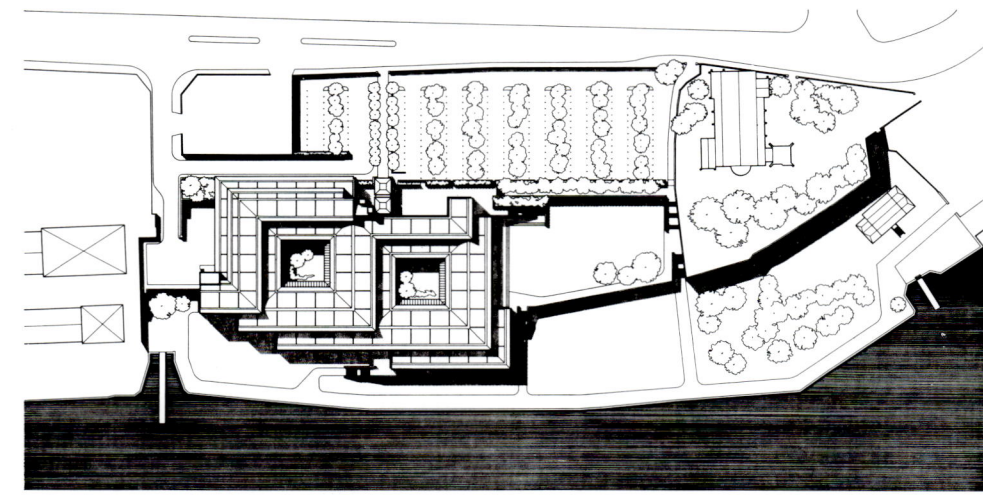

Scale 1:500

1974年，伦敦劳埃德保险公司决定建造一座新的大楼以迁入一些部门。新建筑所在之处曾是查莎姆海军造船厂的一部分，位于梅德威河畔，远眺还可以看到宏伟的罗彻斯特城堡。

规划要求设计一座20,000m²的建筑，内有办公室、信息中心及附属设施（包括一间员工餐厅）。如何利用场地的特征十分重要，尤其是场地前方正对的河流及远处的景观。

设计规划了两个庭院，建筑围绕庭院布置。建筑物利用了场地原有的坡度，从而减小了建筑物的体量。

建筑物中有4层是办公室，可以根据部门的需要调整空间设置。除了计算机房为了保证设备最佳的安全性而采用空调系统和封

1　Site plan
2　The building overlooks the River Medway
3　The lower courtyard; external timber louvres filter sunlight into the main staircase and overhanging eaves protect offices from the summer sun
4　The lower courtyard; external timber louvres filter sunlight into the main staircase and overhanging eaves protect the offices from the summer sun

1　场地平面图
2　从建筑物俯瞰梅德威河
3　建筑外立面
4　下方的庭院，窗户采用外挂的窗檐以防护夏日的阳光

3

4

Lloyd's of London Administrative Headquarters 75

sited to provide maximum security, working areas are arranged to give occupants good outlook; natural ventilation is provided by opening windows which are shielded from the summer sun by overhanging eaves.

Pitched roofs are used to create a building whose profile is appropriate to the setting. The materials used—grey tiles for the roofs and red brick for the walls—have also been chosen to ensure that this building fits into its surroundings.

闭设计以外，工作区都采用可为使用者提供开阔视野的设计，窗户可以开启以进行自然通风。为了防止夏天的日晒，窗户也采用了外挂的窗檐进行保护。

屋顶倾斜找坡，使得建筑物的轮廓与周边环境更为协调。材料选择的原则，如屋顶选用灰瓦和外墙选用红砖，也是要保证建筑与周边环境相适合。

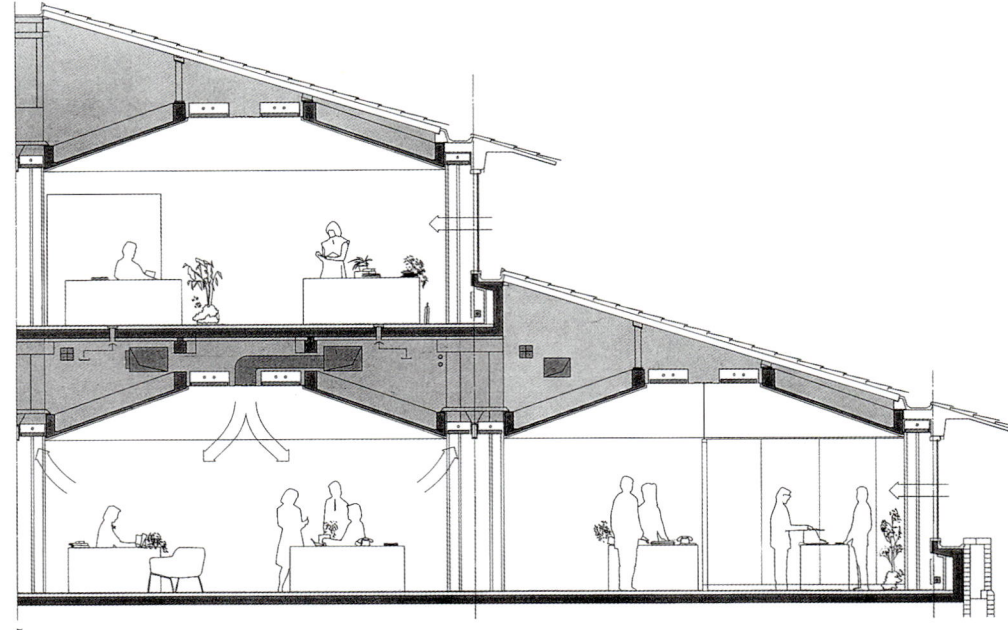

5

6

5　The section, stepped to follow the slope of the site, consists of a square module defined by structural pyramids with clusters of columns at each corner
6　View of entrance hall staircase showing the relationship of timber panelling, precast concrete, carpet, stainless steel and quarry tiles
7　Detail of column
8–9　The entrance hall steps down the site with views to the river

5　剖面图，建筑的阶梯式设置与场地的坡度相合，结构由方形模块组成，角部设置集束柱
6　入口门厅和楼梯内景，可以表现木制镶板，预制混凝土，毡毯，不锈钢，大铺地砖等材料之间的关系
7　角柱构造细部
8–9　入口门厅楼梯，由此可远眺楼前的梅德威河

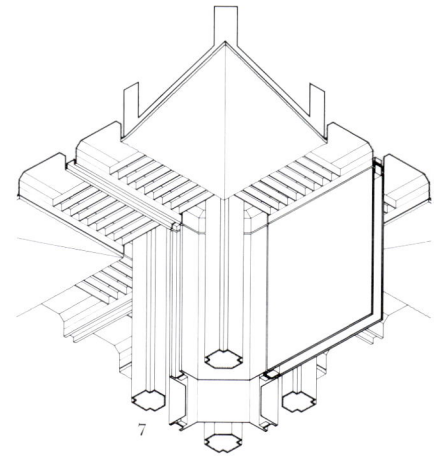

Lloyd's of London Administrative Headquarters 77

CEGB South West Region Headquarters

Design/Completion 1973/1978
Bedminster Down, Bristol
Central Electricity Generating Board
24,000 square metres
Concrete, precast floor planks
Slate roofs, hardwood, masonry walling

英国中央电力局西南区总部

设计/完成　　1973/1978
北德明斯特高地，布里斯托尔
英国中央电力局
24,000m²
混凝土，预制木地板
石板屋顶，硬木，砌块墙体

The headquarters building provides a total of 24,000 square metres of space dedicated to offices, laboratories and service areas for 1,200 staff. It is situated on a 7-hectare site on Bedminster Down, and commands fine views of open countryside, the city of Bristol and the Avon Gorge.

The building has been designed to minimise its impact on the surrounding landscape and is planned around an internal pedestrian "street" containing all communal facilities.

Open timber trusses support pitched roofs with high level glazing which provides good levels of natural daylight. Deep overhangs and ventilated roof cavities help to reduce solar gain.

The supply air to the occupied spaces passes through hollow core precast floor planks. On summer nights the ventilation system passes outside air through the cores to cool the structure. During the day the supply air is cooled during its passage

Continued

　　总部大楼建筑面积24,000m²，包括办公室，实验室和为1200名员工配备的服务设施。建筑坐落在北德明斯特高地的一块7hm²的场地之中，可以俯瞰到开阔的乡村地带、布里斯托尔城和埃文河谷（Avon Gorge）。

　　建筑设计的原则是将其对周围景观的影响最小化，建筑物围绕一条内部的人行"街道"而建，所有的公共设施就设在"街道"之中。

　　屋顶倾斜找坡，采用敞开的木制桁架支撑，采用高窗可以获得很好的自然光线。厚重的悬挑屋顶和通风管孔可以有效地减少日照热量。

　　通风系统设在空心的预制木地板之中。在夏季，晚上户外空气流过通风系统可以降低室温；而在白天空气进入建筑物时，在通过木地板时就会得到有效的冷却，这样的冷

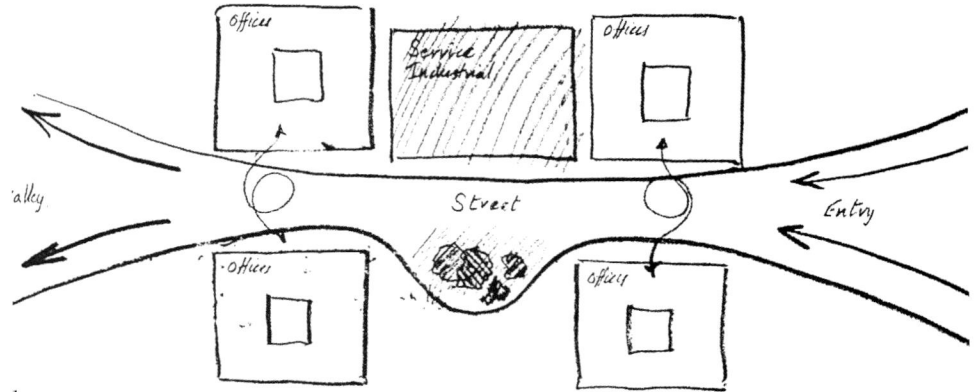

1

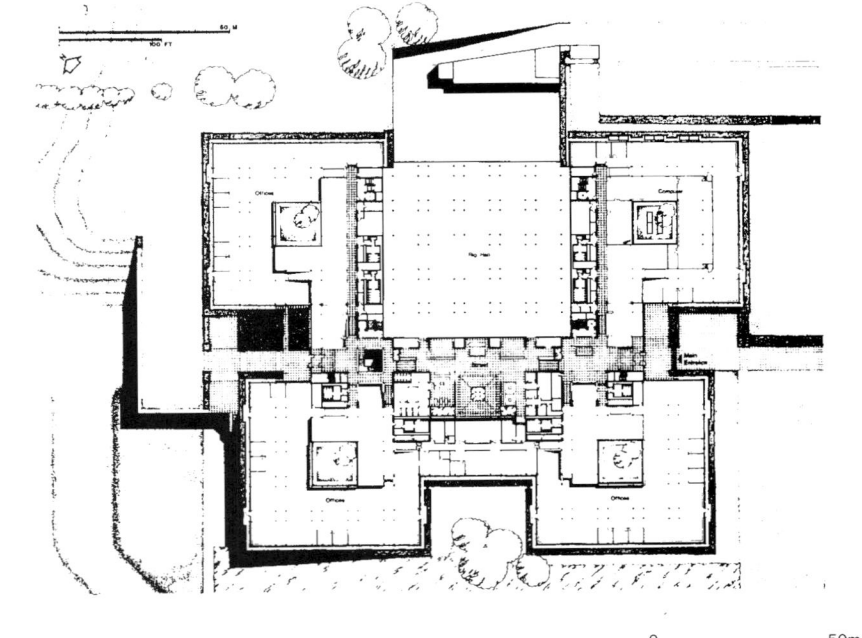

2

1　The offices create a series of courtyards linked by a street which has views out over the countryside
2　Floor plan
3　Section
4　Interior corner detail
5　Protecting eaves, dwarf walls and planting extend the horizontal forms of the building and tie them into the landscape

1　办公楼中设置了一系列的庭院，由一条街道相连，可以俯瞰楼外的乡村景色
2　建筑平面图
3　剖面图
4　内部转角处细部
5　长挑的屋檐，矮墙和植物拓展了建筑物的水平体型，使之与周围景观相协调

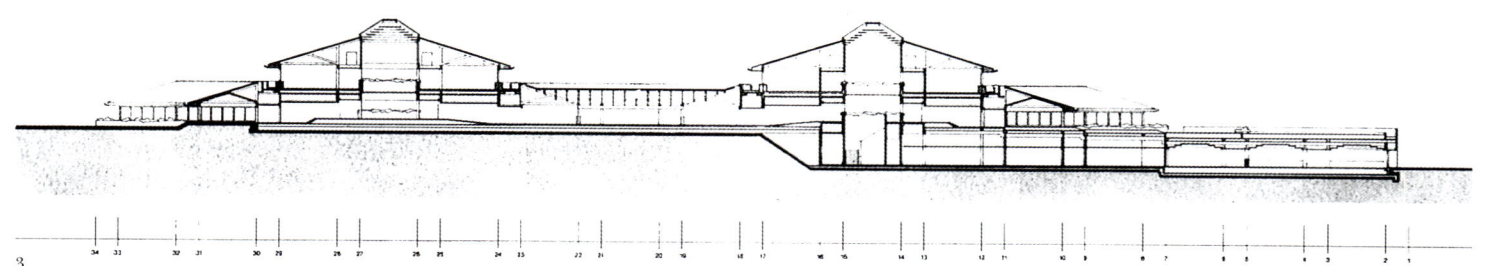

CEGB South West Region Headquarters 79

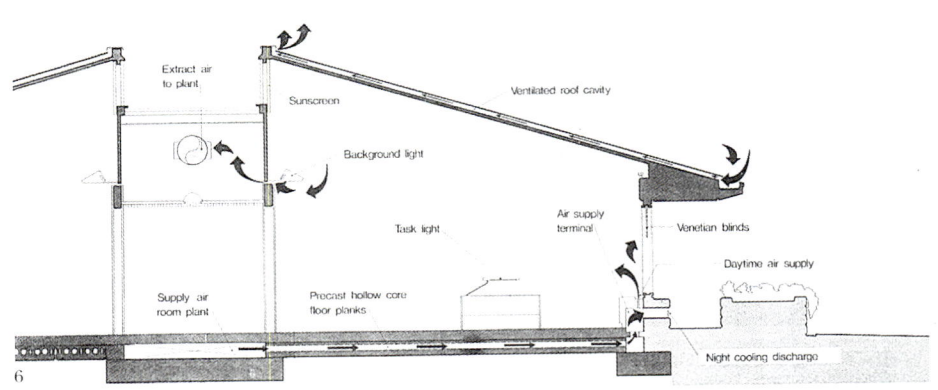

through the cores; this cooling is generally sufficient to maintain comfortable temperatures. This system takes maximum advantage of natural temperature swings, reducing the dependence on mechanical refrigeration for space temperature control.

却方式足以有效地保证室内合适的温度。这一系统最大程度地利用了自然温差的作用，减少了对机械制冷的依赖。

6 Section through an office
7 Special test rig bays
8 Main staircase handrail detail
9 Research laboratories
10 To save energy the offices are naturally lit by large perimeter windows sheltered from direct sun by deep eaves
11 Exposed laminated timber trusses and clerestorey windows to office areas

6 某办公室剖面图
7 安放试验台的开间进行了特别的设计
8 主楼梯扶手细部
9 实验室
10 办公室采用大开窗自然采光以节约照明用能源，为避免直接日照，使用长挑的屋檐遮挡
11 办公区域中无遮蔽的叠层式木桁架和天窗

11

Selected and Current Works

作品精选

1980s

Factory for Trebor Limited

Design/Completion 1977/1980
Colchester, Essex
Trebor Limited
9,400 square metres
Steel, precast concrete
Facing brick, exposed steel, glass

Trebor Limited commissioned Arup Associates to design a new factory complex which was to be light and spacious with a high standard of physical and environmental amenities for all employees.

The project comprises a series of connected buildings with landscaped areas between, linked by a circulation and services "street". Two building types predominate: those which require height —either for storage racks, sugar silos or specialist equipment—and those which do not. Buildings in the former category have few windows and are simply expressed in facing brick and exposed steel. The lower buildings are glazed and are generally air-conditioned.

In these areas the structure consists of steel columns which support primary beams with pairs of secondary beams above.

Continued

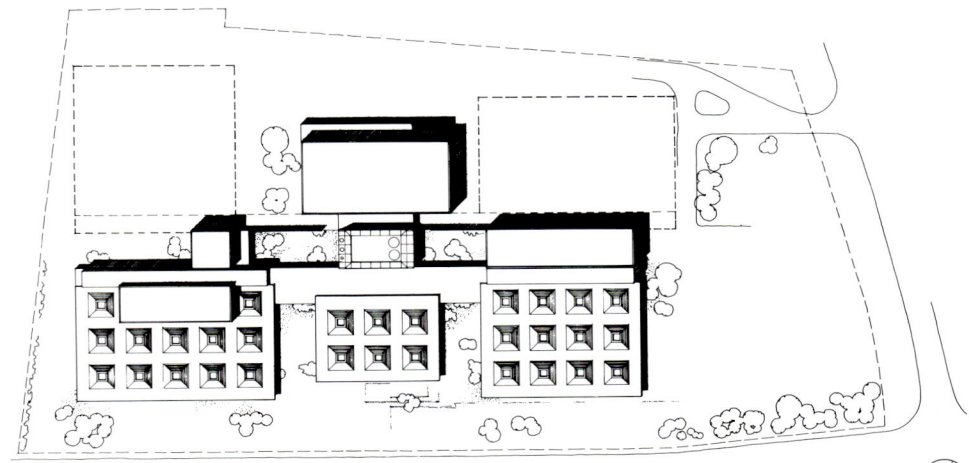

Scale 1:500

1 Site plan
2 The new factory occupies a site close to the countryside on the edge of Colchester
3 Exterior of the buildings seen through trees

3

4

Precast concrete was used for the roof structure, forming a pyramid in each structural bay, with a large rooflight at the centre. Horizontal runs for ducts and services are within the secondary beam zone, with vertical drops housed within the flanges of each column.

屋盖采用锥形结构，采用预制混凝土，以开间为基本模数。锥结构的中心布置大型的屋顶照明设施。水平方向的设备管线布置在次梁位置，垂直方向的线路则嵌入在柱的凸缘之中。

5

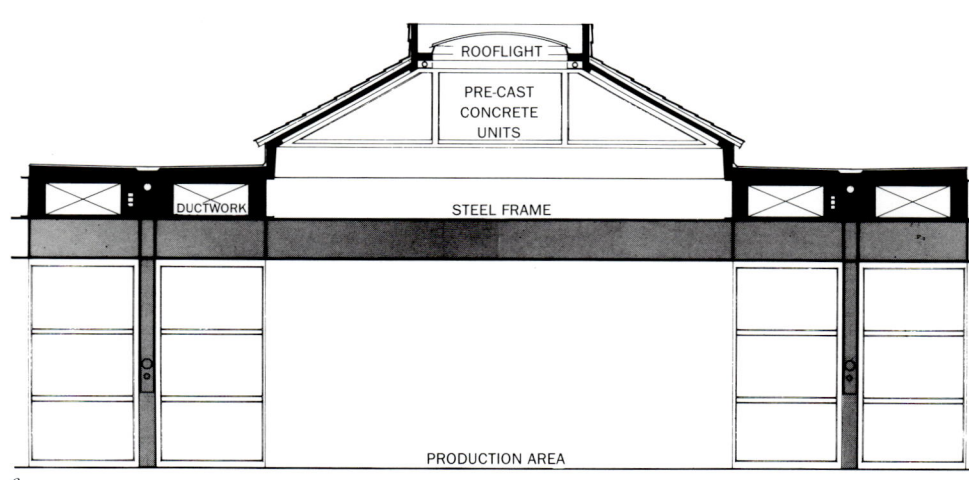

6

4 In the production areas some services are taken down to connection points on the columns
5 The provision of natural light, together with the exposed steel frame, white walls and ceilings, creates a good working environment
6 Section of the factory
7 Planned around a series of landscaped courtyards, the factory provides a good working environment as well as an efficient setting for sophisticated manufactuing processes
8 Elevation of a typical bay

4 生产区的一些管线出口设在柱的连接节点处
5 自然光线、外露的钢框架结构、白色的墙壁和顶棚，共同组成了良好的工作环境
6 工厂建筑剖面图
7 工厂环绕一系列精致的庭院而建，可提供良好的工作环境，同时也能有效地满足复杂的生产流程的需要
8 标准开间正面图

Factory for Trebor Limited 87

Bedford School

Design/Completion 1979/1981
Bedford, Bedfordshire
Bedford Charity
5,000 square metres
Timber, plaster, Weldon stone

In March 1979 the historic main school building at Bedford School was largely destroyed by fire. The physical focus of the school was the Great Hall which was surrounded by classrooms on three sides. The north elevation of this historic building was designed as a Gothic facade and after the fire the brick shell was all that remained. The new design maintained the Great Hall as the focus, but rebuilt it from the first floor up.

As rebuilt, the Great Hall can be used in an "amphitheatre" form for school assembly or in a conventional length-on manner for musical concerts. The ground floor was replanned, incorporating a new south entrance from the school quadrangle, redesigned circulation and additional accommodation for classrooms and offices. A new third floor was added into the existing roof volume with windows concealed behind the roof pitch.

Continued

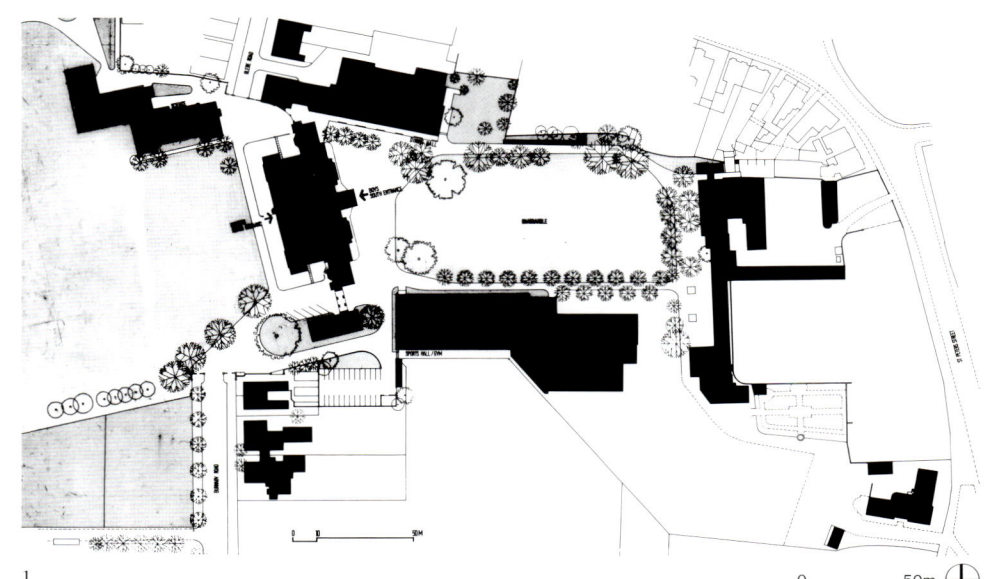

1 Site plan
2 The original school building was destroyed by fire, leaving only the shell
3 A new entrance was created as a part of the rebuilding

3

4   The Great Hall and galleries can accommodate over 1,000 people
5–6 The hall was restored and a new timber roof added

4   大礼堂和楼座可以容纳超过1,000人
5-6 大厅进行了重建，增加了一个新的木结构屋盖

4

The Great Hall, which now has magnificent views through the north stone tracery windows, was reconstructed with timber trusses, timber ceiling and floor. Including the side and end galleries, it can accommodate over 1,000 people. Its roof is supported on a series of interlocking timber vaults constructed in laminated softwood and braced with a boarded ceiling of Western Red Cedar. The doors and panelling are in English Chestnut and the strip floor is Danish Oak. The original Weldon stone tracery windows have been carefully rebuilt and columnettes and plinths added within the Great Hall.

大礼堂采用木桁架结构、木制顶棚和地板，从北面的石制窗格中看去，大厅内一片宏伟华贵的景象。如果将两侧和端部的楼座也计算在内，整个礼堂可以容纳超过1,000人。屋盖采用一系列互锁式的木质拱顶支撑，拱顶材料采用层合软木，顶棚采用西红木，同时为拱顶提供平面支撑。门窗镶板采用英国栗木，条状木地板采用丹麦橡木。原来的韦尔登石制窗格进行了小心谨慎的重建，根据需要还增加了柱和柱基。

5

Babergh District Council Offices

Design/Completion 1977/1982
Hadleigh, Suffolk
Babergh District Council
4,000 square metres
Brick, reinforced concrete floors
Suffolk pantiles, orange brick, timber trusses

The brief for the design of the headquarters for a newly created Local Authority in a small town in Suffolk called for office accommodation for about 200 people working in five departments, as well as a council chamber, associated facilities, a restaurant, main public entrance and waiting areas.

The site, overlooking the River Brett, included a 19th century granary, a number of houses and a cottage. Five of these buildings were of special architectural or historical interest, and these formed the nucleus of the new headquarters.

Continued

巴伯夫区自治会办公楼

设计/完成　1977/1982
哈德雷夫，萨福克郡
巴伯夫区自治会
4,000m²
砖，钢筋混凝土楼板
萨福克波形瓦，橙色砖，木桁架

位于萨福克郡的一座小镇成立了新的地方自治机构，按照要求，自治会总部办公楼应能够容纳约200人在其中工作。大楼按功能分成这5个部分，包括会议室、配套设施区、餐厅、主公共入口和等待区。

从场地可以俯瞰布雷特河的景色，小镇的建筑包括一座19世纪建造的粮仓、一系列住宅和别墅建筑。这些建筑中的五座有着特别的建筑和历史风格，构成新办公楼的核心特色。

1

1 Site plan
2 The new offices are accommodated within a series of new buildings linked to restored historic houses and a granary
3 A glazed bay framed by old and new buildings provides a waiting space for visitors

1 场地平面图
2 新的办公室的风格与一系列新建筑、修复的古建筑和粮仓相协调
3 新老建筑之间采用玻璃框架间隔出开间，作为参观者的等待区域

Babergh District Council Offices 93

The scheme was developed to restore and re-use these existing buildings and link them to new additions which more than doubled the existing floor area. It was designed to respect the existing character, scale and patterns of building in the town. The design develops these existing patterns to establish a clear circulation system within the headquarters and create a series of external landscaped courtyards enclosed by buildings both new and old. The public entrance and the new council chamber are the hub of a series of significant routes around which the complex is organised.

设计方案将这些已有建筑重新修复和重新利用，并且将其与新增的建筑区域相连接，使原来的建筑平面面积扩大了两倍以上。设计较好地考虑了小镇原有建筑的风格、规模和样式。设计还发展了原有的建筑样式，在新的办公楼中建造了良好的供应流通设施，在建筑外部，新老建筑的围绕中设计了精美的庭院。公共入口和新会议室设在建筑内主要通道的交汇处。

4

5

4　一座已有建筑上层楼面的改造
5　改造后的粮仓底层用作办公室
6　从会议室的露台上望去，可以看到被建筑环绕的庭院和远处的乡村地带

4　Upper floor conversion of one of the existing buildings
5　Ground floor offices in the converted granary
6　The terrace off a council meeting room looks onto an enclosed courtyard and the countryside beyond

7　View from reception to the glazed waiting area with the converted granary beyond
8　The council chamber

7　从接待区看到的玻璃墙后的等候区，远处为改造过的粮仓
8　会议室

8

Babergh District Council Offices 97

Gateway 2

Design/Completion 1981/1982
Basingstoke, Hampshire
Wiggins Teape (UK) PLC
22,000 square metres
Steel, precast concrete
Glass

盖特威 2 号楼

设计/完成　　1981/1982
贝辛斯托克，汉普郡
维金斯蒂普（Wiggins Teape）（英国）公司
22,000m²
钢，预制混凝土
玻璃

Gateway 2 was designed to satisfy a developer's brief to provide uniform office width, at low cost and with natural ventilation.

The building, with a floor area of about 16,000 square metres, was planned around a central, top-lit atrium; office accommodation has views to the atrium as well as to the outside. The ceilings in the office areas are formed from structural units which incorporate light fittings with acoustic absorbers. Service voids are located above the ceilings.

The atrium is a major amenity which also acts as a natural chimney, drawing fresh air through the offices and out vents in the roof. This makes it possible to ventilate naturally a building which would normally have required air-conditioning. The occupants can control the degree of ventilation by opening or closing windows.

Continued

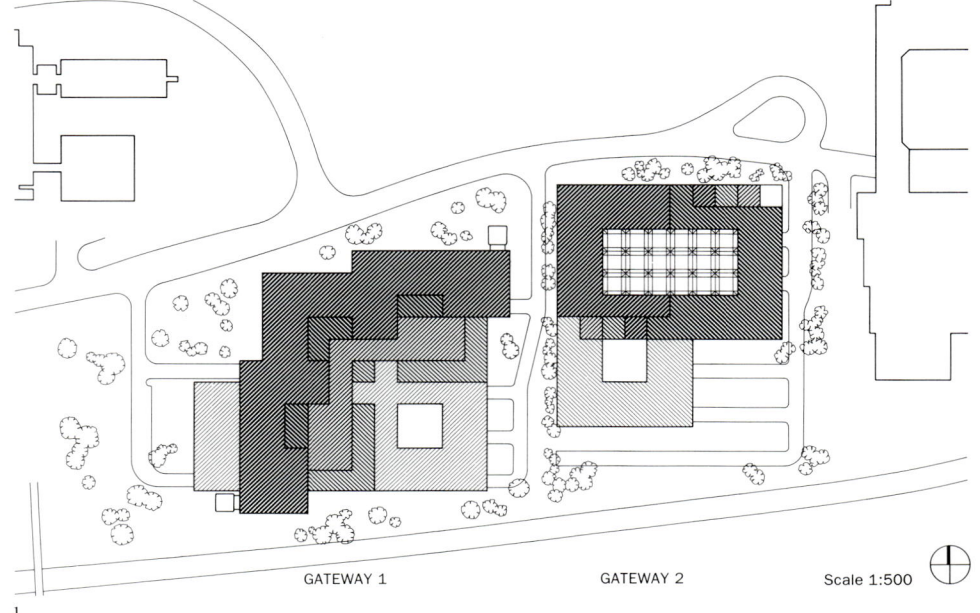

1

项目规划要求建筑能够提供均匀的办公室空间，采用自然通风，以符合经济节能的原则。

大楼建筑平面面积约 16,000m²，设计了采用顶部采光的中心庭院，从办公室里可以眺望建筑外部景观和中心庭院的景色。办公区域的顶棚由结构单元组成，单元内含照明设施和吸声装置。供应管线设置在顶棚上部。

中心庭院是主要的休闲设施，同时也起着自然通风口的作用。新鲜空气从庭院顶部的通风口抽入办公楼内。这使原本通常采用空调设备的建筑可以自然地实现通风。使用者可以通过开关窗户控制自然通风的程度。

2

1　Site plan
2　The structural frame is expressed externally to provide sun shading and a series of stepped, landscaped terraces
3　The atrium

1　场地平面图
2　结构框架展现在外部，可以遮蔽阳光，并营造一系列阶形露台
3　中心庭院

Circulation galleries cross the atrium, providing direct access between the offices and the glass elevators. Whilst satisfying primarily functional needs, timber screens and a tracery of steelwork and sunshades are also designed to scatter the light and enrich the central space. At ground level the atrium is used as a meeting place, a badminton court and, on occasions, a place for social gatherings.

廊道穿过中心庭院，将办公室和电梯间直接相连。为了满足建筑的功能需要，在中央庭院四周设置了木制屏障、钢制窗格和天幕以削弱阳光的影响，同时也起到装饰作用。中心庭院地面可以用作会议场地、羽毛球场，或者偶尔用作社交集会区域。

4

4　Windows for the offices are designed within a timber framework
5　The atrium encourages natural ventilation within the building
6　Seating and waiting spaces within the atrium
7　The atrium provides daylight to the offices and creates an amenity at the heart of the building

4　办公楼的窗户设计采用了木制窗框
5　中心庭院促进了建筑物的自然通风
6　中心庭院内的座位和等待区域
7　中心庭院可以为办公室提供自然光照，且可用作建筑内部的休闲场所

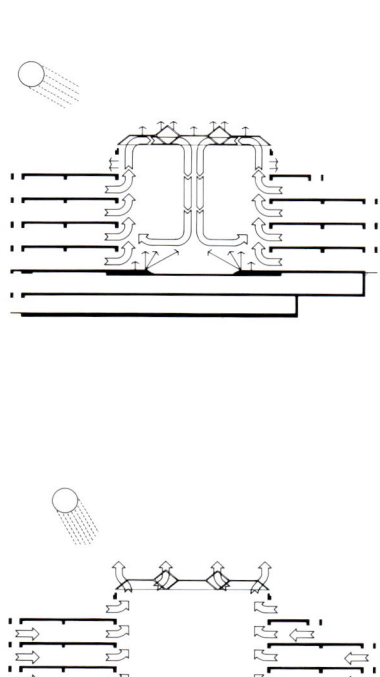

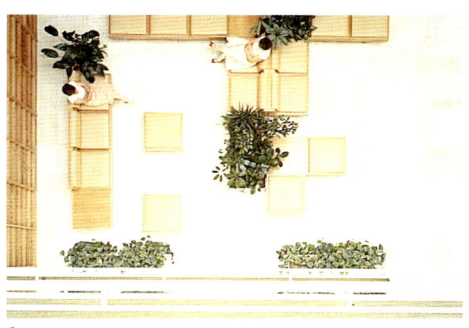

8　The central atrium
9　Sunscreens to the rooflights within the atrium also act as accessways for maintenance

8　中心庭院
9　中心庭院顶部的遮光天幕

8

Briarcliff House

Design/Completion 1978/1983
Farnborough, Hampshire
Imperial Group Pension Trust Limited
9,400 square metres
In situ concrete
Bronze tinted glass, aluminium mullions

布莱克里夫大楼

设计/完成　1978/1983
法伯鲁夫，汉普郡
帝国养老金信托集团有限公司
9,400 m²
现浇混凝土
铜色玻璃，铝合金框

The 0.4-hectare site is situated at the southern end of a shopping complex in the centre of Farnborough. It is bounded on the south, east and west by roads and is located close to the Royal Aircraft Establishment.

The brief called for an office building of approximately 9,300 square metres to accommodate some 400 staff, together with an extension of the existing Kingsmead Shopping Centre Mall at ground level. In addition to offices, accommodation was to include a computer suite, staff dining room, bar, executive dining areas, conference facilities, printing department and other associated functions. Special consideration had to be given to solar gain and the problems of traffic and aircraft noise on the site.

The building is planned in the form of a "U" which follows the southern site boundaries. At its centre is a new landscaped roof garden covering the

Continued

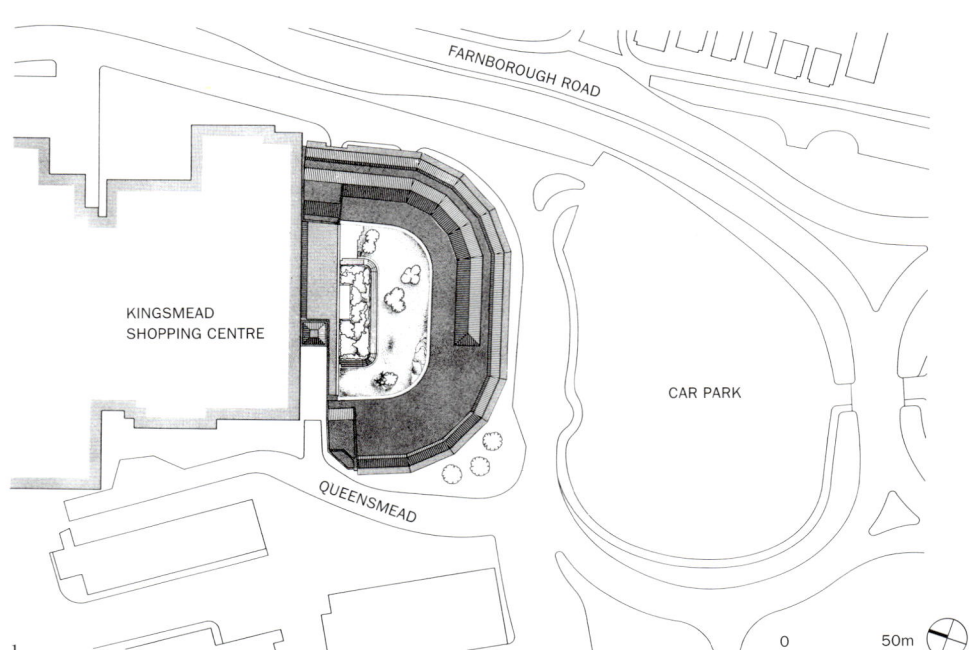

该项目占地 0.4 hm²，位于法伯鲁夫市中心商业区的南端。场地东面、西面和南面都被道路包围，毗邻皇家空军基地。

计划要求建造一幢约 9,300 m² 的办公大楼，以容纳约 400 人在其中工作。建筑在地面层还将用于扩展原有的金斯麦德购物中心。除办公区外，大楼内还包括计算机设备房、员工餐厅、酒吧、管理层用餐区域、会议室、印刷室和其他附属设施。必须对当地的日照条件、交通问题以及飞机噪声问题加以特别的考虑。

建筑物根据了场地三面环路的边界形状，采用了"U"形的建筑形式。在一层商业区的中心精心设计了一座屋顶花园。

1 Site plan
2 The external glass wall, designed as a double skin, reduces solar gain and traffic noise on a prominent urban site
3 The staff restaurant looks onto a sheltered landscaped garden
4 A glazed canopy provides a covered walkway along the street and marks the main entrance to the offices

1 场地平面图
2 外部的玻璃幕墙，采用了双层设计，可以减弱市中心场地日照和交通噪音的影响
3 从员工餐厅内透过遮帘可以看到精心设计的花园
4 在毗邻建筑物的街道人行道架设玻璃天棚，指引出办公楼的主入口

shops at ground level. The main staircase and lifts are located at the centre, with escape stairs and toilets at the end of each wing. Between these service cores on the upper floors 14.4 metre office spaces provide discrete areas of approximately 800 square metres. The main horizontal circulation overlooks the roof garden. Enclosed workspaces can be located on the southern side of the offices.

The southern elevation is designed as a double skin to minimise solar gain and traffic noise. The outer skin, which is totally glazed, acts as a fly sheet, shielding the building from the weather and providing consistency of appearance; the inner skin is made up of solid and glazed panels which can be arranged to suit changing needs. Heat in the space between the skins can be used in the winter but is prevented from entering the offices in the summer. The primary air handling units are accommodated at roof level with the main air supply ducts feeding down between the skins.

上部楼层 14.4m 高，每层约 800m²。主要的楼梯和电梯间设置在中心位置，紧急疏散楼梯和卫生间设置在两侧端部，办公区则间隔布置在这些服务设施之间。环绕中心的水平走道上可以俯瞰屋顶花园。封闭型办公室可设在办公层的南面。

建筑南立面设计了双层的幕墙，可以减小日照和交通噪声的影响。外层采用完全的玻璃，作为建筑与外部环境之间的屏障，且可营造统一的外观效果；内层采用实体材料和玻璃面板，可以通过合理配置适应不同的需要。两层幕墙之间的热量在冬天能够加以利用，在夏天则无法侵入建筑内部。主要的通风设备安置在屋顶之中，供风管道主要设在双层幕墙之间。

5

| 5 | The main entrance to the office building, showing the first-floor and restaurant beyond |
| 6 | Each elevation responds to its orientation: the offices (left), have automatic external drop blinds only over the glazing panels that catch the sun; the restaurant (right), which faces south into the garden court, has fixed louvres |

5 办公楼的主入口，可以看到建筑一楼和稍远的餐厅
6 立面设计与其方向相关：办公室（左），仅在玻璃幕墙外设置了自动升降垂幕；餐厅（右），面向南面的花园，则设置了固定的散热窗

International Garden Festival Hall

Design/Completion 1982/1984
Liverpool
Merseyside Development Corporation
7,500 square metres
Profiled aluminium sheet, polycarbonate sheet

The Festival Hall was the winning entry in a national architectural competition to provide a focus for the Garden Festival held in Liverpool.

The structure of the building provides its form. The weathering envelope is composed of two related forms: the dome and the vault. The dome is halved and then rejoined by the linear vault. The structural and economic efficiency of this curvilinear form results in a low material content and a favourable ratio between surface and plan area.

The building was designed to provide a 7,500 square metre column-free enclosure to provide maximum flexibility of use for the Garden Festival and to allow for later use as a sports complex.

Continued

国际园艺节展厅（International Garden Festival Hall）

设计/完成　1982/1984
利物浦
默西塞德郡开发公司
7,500m²
异型铝合金板材，聚碳酸酯板材

该展厅的设计是在为利物浦国际园艺节展厅举办的全国性建筑设计竞赛中脱颖而出的。

设计利用结构来塑造建筑的体型。围护外壳由两种相关的形式组成：穹顶和拱顶。圆形屋顶中间分开，由线性的拱顶重新联为一体。这样的曲线形式的结构和经济效益可以节约材料，且获得较好的表面/平面比率。

展厅可以提供 7,500m² 的无柱室内场地，能够为园艺节的空间使用营造最大的灵活性，园艺节以后这里还可以作为综合运动场馆使用。

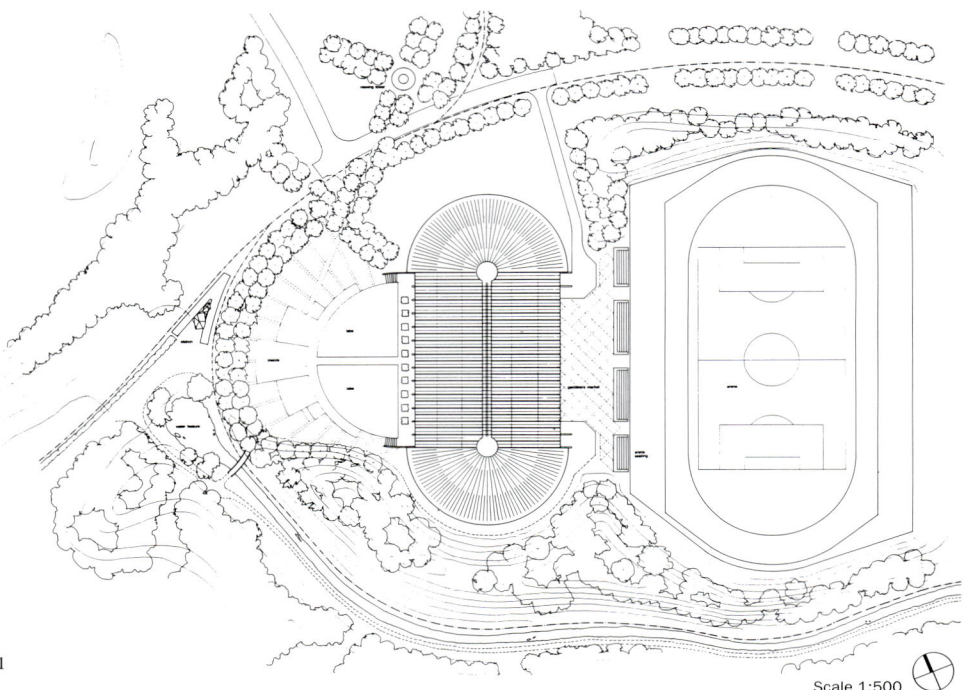

1
Scale 1:500

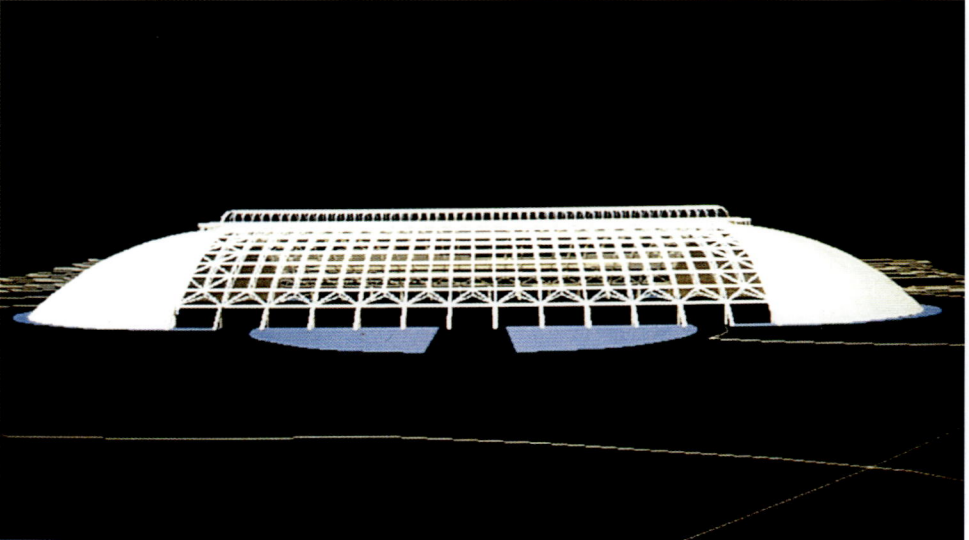

2

3

1   Site plan
2–3 Computer drawings showing the overall
    three-dimensional form of the building

1   场地平面图
2–3 计算机仿真图，展现建筑的全息三维景象

International Garden Festival Hall   109

The envelope is energy-efficient and has low maintenance costs. The long, low dome structures are clad in profiled aluminium sheet with a composite base to provide a high level of heat insulation, acoustic treatment and a vapour barrier that will withstand high humidity levels. The central transparent vault is covered in 20 millimetre polycarbonate sheet treated to combat glare by refraction within its cellular structure. This creates an environment which is suitable for plants, but which will also be appropriate for sports and leisure activities.

The reflective "glassy" finish provides an appropriate sparkle to the Festival Hall, in keeping with the great glass structures of Paxton and Decimus Burton.

围护拱壳符合节能的要求，维护费用也比较经济。较矮的穹顶结构采用异型铝板盖层和复合材料基础，能提供很好的隔热效果、声音处理和防水效果。中间的半透明拱顶采用20毫米厚的聚碳酸酯板材覆盖，这种板的蜂窝结构产生的折射效果削弱了直接日照的光线强度。这样的光照环境适合于植物的生长，也适合运动和休闲活动的开展。

展厅外部采用了反光的玻璃状面层，使建筑呈现出一片光芒闪耀的华丽景象，这与帕克斯顿、德西姆斯·伯顿等人建造的大型玻璃结构有共通之处。

4

4　The hall was designed to provide large, column-free spaces to accommodate exhibitions, events and large displays
5　Elevation
6　The Garden Festival Hall was the focus of an extensive development on a large site beside the River Mersey

4　展厅设计要求提供宽阔的无柱空间，以满足展览、活动及大型展示的要求
5　立面图
6　园艺节展厅是墨济河畔大型开发区的核心建筑

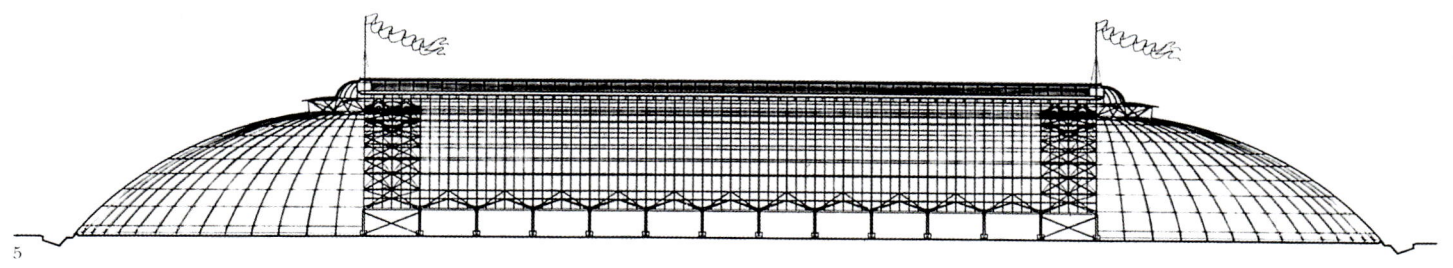

International Garden Festival Hall 111

7 The form of the building recalls the garden buildings of Richard Turner and Decimus Burton
8 The view of the ceiling, a painting by Ben Johnson, shows the sky-lit apsidal end of the roof

7 建筑的形式追随了理查德·特纳，德西姆斯·伯顿等人的花园建筑风格
8 本·约翰逊绘制的顶棚视图，拱顶开设天窗进行自然采光

7

International Garden Festival Hall 113

1 Finsbury Avenue

Design/Completion 1982/1988
London EC2
Rosehaugh Greycoat Estates Limited
50,346 square metres
Steel frame with metal decking, concrete floors
Bronze anodised aluminium curtain wall and sunscreens, glass

芬斯伯里大道 1 号（1 Finsbury Avenue）

设计/完成　　1982/1988
伦敦东部中央 2 区
洛斯豪夫·格雷可特（Rosehaugh Greycoat）房地产开发有限公司
50,346m²
钢框架结构，金属外装，混凝土楼板
镀铜铝合金幕墙及遮光篷，玻璃

This three-phase urban development project provides about 50,000 square metres of offices, as well as shops, restaurants and a sports centre.
The first phase contains 25,000 square metres of rentable office space with other uses.

Eight storeys high, with stepped back landscaped terraces at the fifth and sixth floor levels, the building is planned around a full height central atrium capped with a large glazed dome.
Two separate circulation cores give access to office areas of more than 3,000 square metres per floor. The building is designed to benefit from natural daylight, while being economical in its use of energy. External and internal shading devices protect the building from the effects of solar gain.

Continued

1　Site plan

Scale 1:200

这一大型城市开发项目分三期建造，计划包含 50,000m² 的办公楼、商店、餐馆和一座运动中心。一期工程为供出租或其他用途的写字楼，建筑面积 25,000m²。

建筑物地上八层，在第五层和第六层设有阶形的观景露台。建筑围绕中央的天井而建，天井高度与建筑等同，顶部设有大型的玻璃穹顶。每层办公室面积大约 3,000m²，垂直交通设在两个单筒中，供办公者上下进出。建筑设计利用自然光照，使其能满足节能要求。外部和内部都设有遮光设施，可以控制建筑的日照效果。

1　Site plan
2　The building on Wilson Street
3　Detail showing the external wall
4　The eight-storey high building, with stepped-back landscaped terraces, defines a new urban square

1　场地平面图
2　威尔逊大街建筑物外景
3　外墙细部图
4　建筑高 8 层，配有内敛的阶梯式露台，营造出新的城市景观

To enhance the existing pedestrian route, a single pedestrian path has been planned between phases 1 and 2, incorporating a square covered by a suspended glass roof. This new intimate space is typical of those traditionally found in the City of London. The newly created square is surrounded by shops, restaurants and pubs.

在一期建筑和二期建筑之间设计了一条简单的人行道,与原来的人行通道一起围成一个小型的庭院,庭院上方还建造了悬挑式的玻璃屋顶。这样一个内部空间形式在伦敦的建筑中颇为典型。庭院四周围绕着商店、餐厅和俱乐部等娱乐场所。

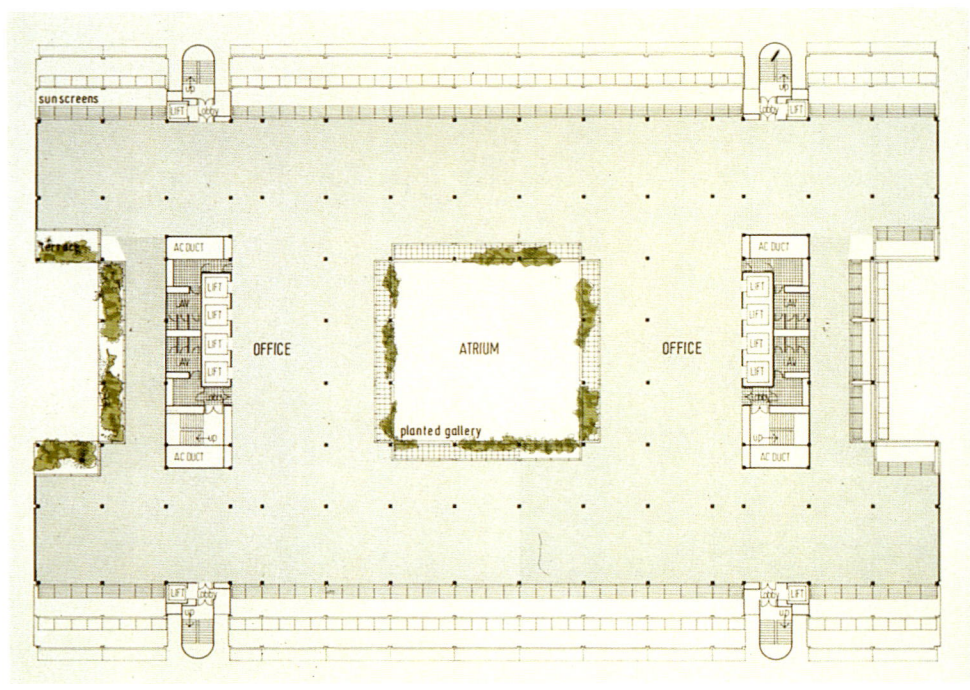

5

5 Seventh-floor plan	5 7层平面图
6 Elevation	6 立面图
7 Section through the development showing the atrium in 1 Finsbury Avenue on the right	7 项目剖面,右边可以看到芬斯伯里大道一期建筑中的中央天井断面

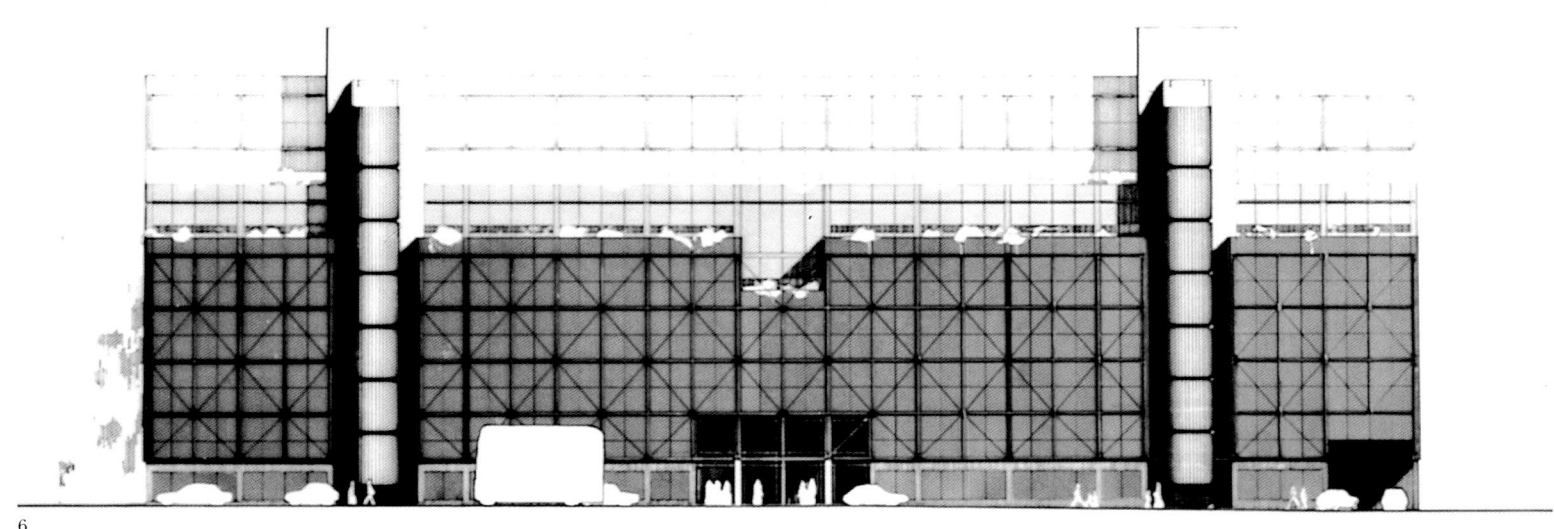

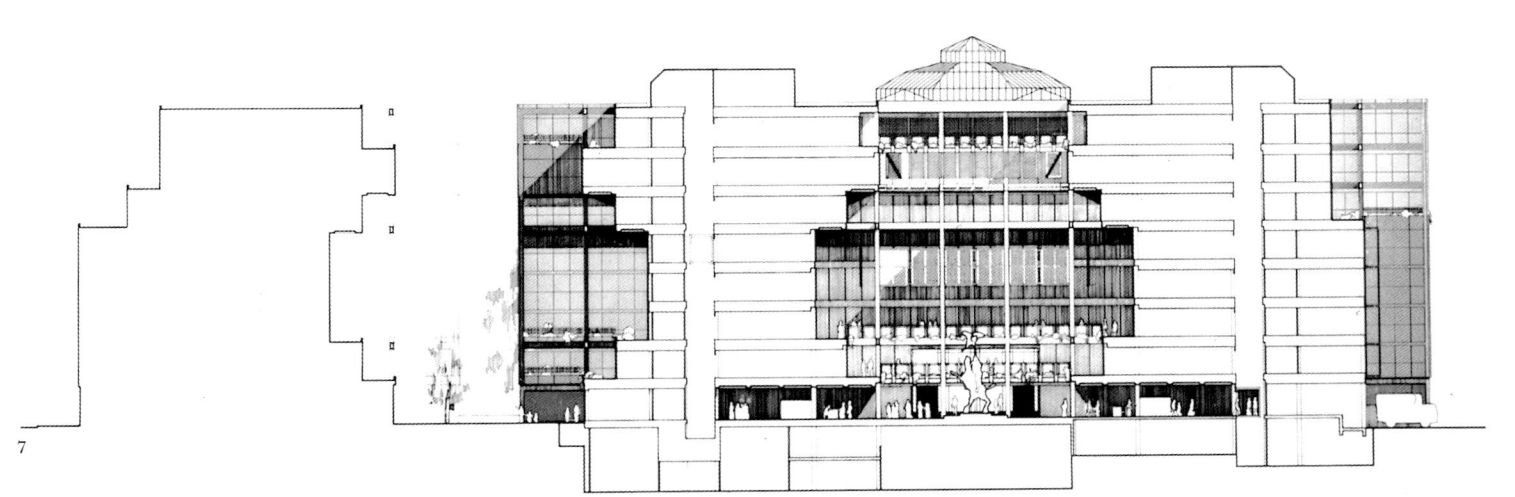

1 Finsbury Avenue 117

8　The building is designed to benefit from natural daylight
9　The sky-lit atrium
10　The atrium, framed by the stepped office floors is shaded by a series of hanging screens

8　建筑设计利用了自然光照
9　庭院采用自然采光
10　办公室面向庭院的一侧设置了一系列悬挂帷幕

8

9

Diplomatic Quarter Sports Club

Design/Completion 1981/1985
Riyadh, Saudi Arabia
Bureau of Foreign Affairs, Saudi Arabia
34,616 square metres
Precast concrete, local limestone walls
Stone-clad concrete blockwork, precast concrete mullions, iroko screens

外交使馆区运动俱乐部（Diplomatic Quarter Sports Club）

设计/完成　1981/1985
利雅得，沙特阿拉伯
外交事务署，沙特阿拉伯
34,616m²
预制混凝土，当地石灰石墙
混凝土预制砌块，石块饰面，预制混凝土门窗棂，栅栏屏障

The sports club serving the community of a new diplomatic quarter is situated on the western side of Riyadh in Saudi Arabia.

The site was a virtually featureless expanse of rocky desert covering an irregular area of some 9.4 hectares. It was located within a new residential neighbourhood with a main pedestrian route designated to run through the site.

The design concept was to create a recreational sports club in an informal, park-like setting. Consequently the sports facilities were planned as a series of separate buildings and enclosures, linked together by pedestrian walkways.

Within this framework the major sports buildings were planned to create two formal planted courtyards, with the sports halls and gymnasium at the eastern end, and the indoor and outdoor pools and the squash courts on the northern boundary. At the hub of the site an existing rise in the ground level was emphasised. The main social facilities of the sports club—the club house and cafeteria—were located here to create a focus for the community.

1

　　该运动俱乐部项目建于沙特阿拉伯利雅得西部一个新建的外交使馆区内，为使馆区内的外国团体服务。

　　场地位于戈壁之中，呈不规则形状，占地约9.4hm²，开阔而没有特色。俱乐部毗邻新建的住宅区，一条人行干道从场地中穿越而过。

　　计划中运动俱乐部应侧重于消遣娱乐功能，因此采用了非正规的公园式规划，运动设施设置在一系列相对独立的建筑或封闭场地之中，彼此由人行通道相连接。

　　在这样的框架下，主要的运动场所规划在两块良好绿化的区域中。运动馆和健身房位于场地的东端，室内和室外游泳池及壁球场位于场地的北部。场地中心原有的高地得到了利用，俱乐部主要的社交场所如俱乐部会所和咖啡室都位于此处，使之成为社区的核心位置。

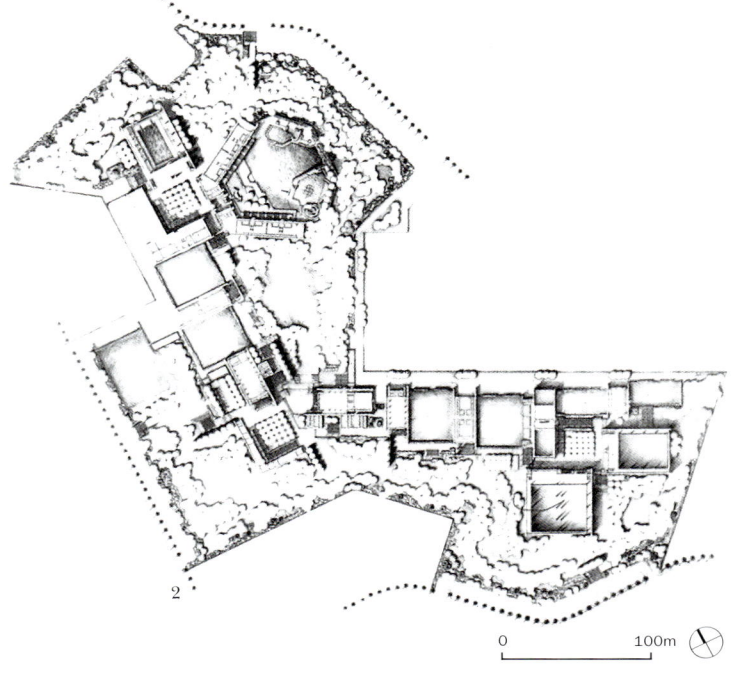

1. Structural beams act as baffles to the light which is then reflected off the walls through iroko slatted screens
2. The club occupies a large site and is planned within an informal, park-like setting
3–4 A series of walls, built in stone, define indoor pavilions for swimming pools, squash courts and gymnasia

1　结构梁起到遮蔽阳光的作用，光线可透过板条栅栏折射到墙壁上
2　俱乐部场地开阔，采用了非正规的公园式规划
3-4　游泳馆、壁球场和健身房的室内空间采用了一系列石砌墙体

Diplomatic Quarter Sports Club　121

Forbes Mellon Library, Clare College, Cambridge

Design/Completion 1981/1986
Cambridge, Cambridgeshire
Clare College
825 square metres
Brick walls, precast concrete and steel
Portland stone panels, clay bricks, lead-coated stainless steel roof with purpose-made clay tiles

福布斯·梅隆图书馆，剑桥克莱尔学院

设计/完成 1981/1986
剑桥，剑桥郡
克莱尔学院
825m²
砖墙，预制混凝土和钢材
波特兰石板，黏土砖，包铅不锈钢屋顶
特别制作的黏土瓦

The brief for this project was to design a library with music facilities to be sited in the existing Memorial Court. The facilities were preferably to be housed within a single building which was to be modern in style and function yet appropriate alongside the existing architecture of Memorial and Thirkill Courts.

On a small site, the building houses the library and sound-proofed music facilities, together with a common room, photocopying room and computer room. The building stands in the centre of Giles Gilbert Scott's original court, but respects the surrounding architecture, complementing rather than intruding on Scott's design. The octagon at the centre of the building is reminiscent of the beautiful 18th century octagonal antechapel in Old Court and, as in the Old Fellows' Library, the books in the new library are arranged against the outside walls, leaving a spacious and well-lit working area at the centre.

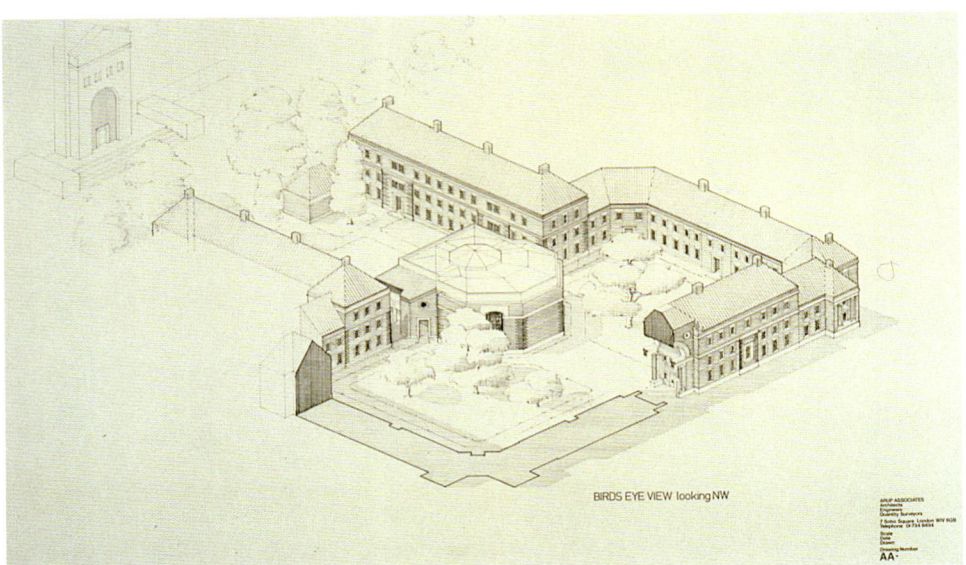
1

2

该项目要求在原有的纪念广场上设计一座配备音乐设施的图书馆。音乐室置于独立的楼内，功能上满足要求，建筑形式现代化同时又与已有的纪念广场周围的古老建筑相适宜。

在不大的场地内，建筑涵盖了图书馆、隔音的音乐室、公共休息室、影印室和计算机房。建筑坐落在古老的圣伊莱斯·吉尔伯特·斯科特广场的中央，由于很好地考虑到与周边建筑的关系，使得它决没有打破斯科特原来设计的格局的感觉，而相反是使其更完美。建筑中心的八角形门厅可以使人追溯到18世纪时老教堂的八角形前庭。和老图书馆一样，新的图书馆中书籍也挨着外墙布置，中央留出充裕而舒适的工作区域。

1　Bird's-eye view looking north-west
2　The new library creates a focus in Memorial Court
3　A seat, set within the bay window, creates a workplace in the library overlooking Memorial Court

1　西北方向鸟瞰图
2　新图书馆成为纪念广场的中心
3　开间窗户旁的座椅可提供工作空间，由此可以俯瞰窗外的纪念广场

3 Forbes Mellon Library, Clare College, Cambridge

4　The west facade of the new library, centrally placed on the axis of the symmetrical Memorial Court, creates new grassed quadrangles
5　The hall is a hub around which the different activities have been planned
6　A top-lit octagonal hall has been formed at the heart of the building
7　The interior of the library

4　新图书馆的西立面，中线与对称的纪念广场的中轴线取齐，营造出新的绿化方庭
5　大厅是建筑物的中心，各种不同功能的房间沿其四周布置
6　八角形的大厅作为建筑物的核心，采用顶部自然采光
7　图书馆内景

4

5

6

7

Forbes Mellon Library, Clare College, Cambridge 125

Broadgate

Design/Completion 1985/1988
London EC2
Rosehaugh Stanhope Developments PLC
149,850 square metres
Red granite, steel frame, glass
Arena: reinforced concrete structure, marble paving

布鲁德门

设计/完成　1985/1988
伦敦东部中央2区
洛斯豪夫·斯坦荷普（Rosehaugh Stanhope）投资开发公司
149,850m^2
红花岗岩，钢框架，玻璃
露天广场：钢筋混凝土结构，大理石地面

Broadgate is a new financial centre located on a 3.5-hectare site which was formerly a car park and railway station, close to the City. The project consists of offices, with shops and restaurants situated around newly created public squares.

These squares, landscaped and planted with semi-mature trees, provide outdoor spaces for a range of activities. During the winter the centre of the square at Broadgate becomes an outdoor ice-skating rink, whilst in summer it is used for concerts, exhibitions and open-air theatre.

Each of the four office buildings is focused around an atrium which brings light and sun into the heart of the building. The external walls of the offices consist of a three-dimensional sculpted grid of red granite which acts as a screen to the windows behind. The buildings step back at the upper floor in order to reduce their impact on the squares. These upper levels incorporate a series of landscaped terraces alongside the offices.

布鲁德门是一座新的金融中心。场地占地3.5hm^2，原为市中心的停车场和火车站。项目包括写字楼、公共广场及广场周围的餐厅和商店。

广场进行了精致的景观设计，种植了生长期的树木，可以为许多活动提供屋外场地。冬季布鲁德门广场的中心地带可以用作为室外滑冰场，在夏天则可以用于举行音乐会、展览和户外演出。

4座写字楼都设计采用了中央天井，目的是为建筑内部提供自然的光照。写字楼的外墙采用立体雕刻的花岗石栅格，对墙后的窗户起到遮蔽作用。建筑物在上层内敛以减少它们对广场空间的影响。这些内敛上部楼层同时也组成了一系列毗连办公室的观景露台。

1

1　The Liverpool Street frontage showing projecting sunscreen in pink granite
2　The marble-lined square at the centre of Broadgate
3　Meticulous detailing and high quality natural materials create attractive external areas designed to weather gracefully
4　A new square has been formed at the centre of Broadgate which is used as an outdoor ice-skating rink in winter

1　建筑面朝利物浦大道，外凸的红色花岗岩对窗户起遮蔽作用
2　布鲁德门中心广场，采用大理石衬砌
3　精致的细部设计和高品质的自然材料，营造了美观且适合环境的外景
4　布鲁德门的中心广场在冬季可用作室外滑冰场

2

3

4

Paternoster Square

Competition 1987
Paternoster Square, London EC1
Paternoster Properties NV
93,000 square metres
Reinforced concrete frame
Glazing, stone and bronze cladding

主祷广场

设计竞赛时间　　1987
主祷广场，伦敦东部中央1区
帕特诺斯特投资开发公司
93,000m²
钢筋混凝土框架结构
玻璃，石料，铜包层

In 1987 five international practices were invited to participate in a competition to prepare a proposal for a large site alongside St Paul's Cathedral in the City of London. Arup Associates were the winners and were appointed in October 1987 to prepare a master plan. The site, known as Paternoster Square because of its historical links with and proximity to St Paul's Cathedral, had been redeveloped after the damage of World War II; however, the office blocks were now considered inadequate for current requirements and were felt to be inappropriately sited. Not only were the blocks unrelated to the cathedral, but the shops were of a poor quality and the public spaces on the site were bleak and uninviting.

The master plan established a framework of streets and squares that linked into the existing city and also created a series of sites for new buildings. These buildings were planned to accommodate a range of uses including offices, a hotel, a new museum and facilities for visitors. The heights of buildings were planned to provide views of St Paul's Cathedral.

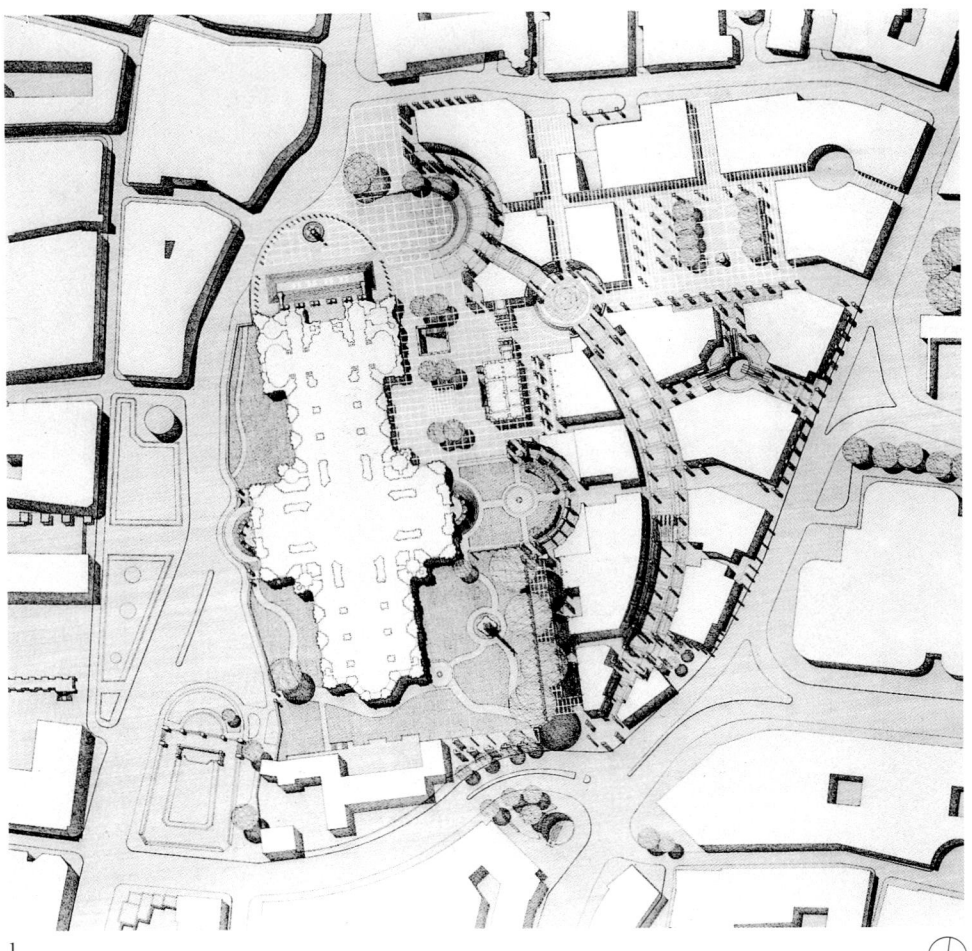

1

Scale 1:30,000

1987年，为开发伦敦圣保罗大教堂旁边的一块大面积场地，开发商邀请了五家国际性的设计公司参加为此举行的方案设计竞赛。阿鲁普联合事务所最终脱颖而出，并在1987年10月被委托进行总规划。主祷广场的名称来自于历史渊源及其与圣保罗大教堂的关系，在二战以后的重建中，这个广场也曾得到过修复。但是，广场周边的办公大楼已不能适合当时的需要，而且其布局也不协调。除此以外，周边的商店也显得十分破败，场地的公共环境让人感觉十分凄凉而没有吸引力。

总规划对原有的街道和广场重新作了规划和布局，开辟了一系列新场地建造新的建筑。这些建筑赋予不同的用途，包括写字楼、一间酒店、一间新的博物馆和参观配套设施。建筑物的高度进行了控制以保持圣保罗大教堂的外景。

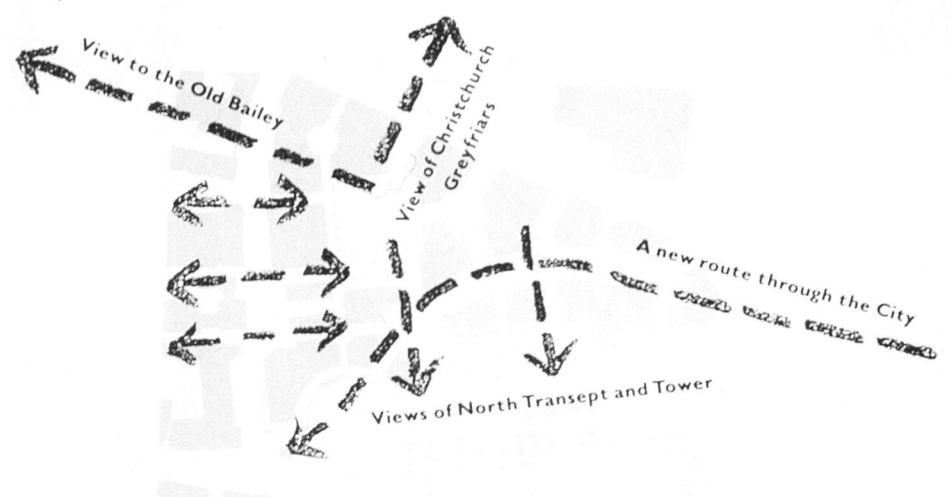

2

1 总平面图
2 重要道路平面图
3 大教堂西面近景
4 模型鸟瞰图
5 广场新建筑剖面图

1 Master plan
2 Strategic routes
3 Cathedral Close, looking west
4 Aerial view of model
5 Section showing the new building on the square

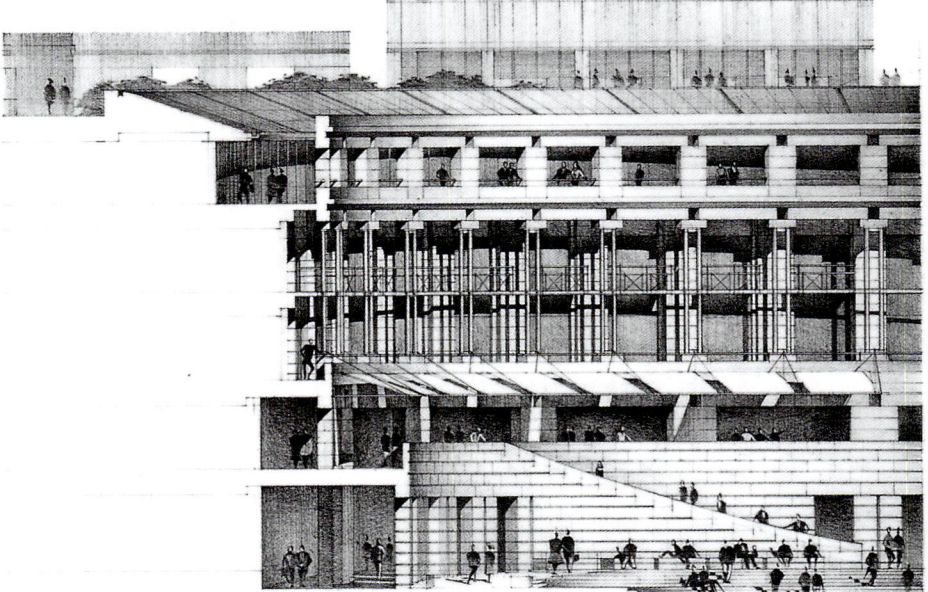

Paternoster Square

6–9 Section through the site showing the new square and west frontage of St Paul's

6-9 场地剖面图,展现新建的广场和圣保罗大教堂西立面

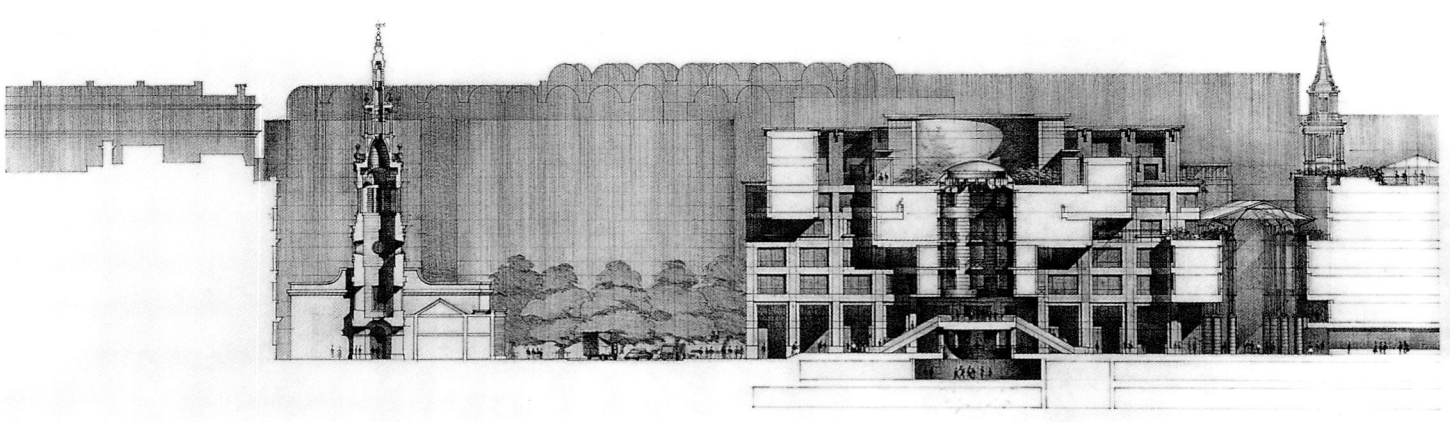

6

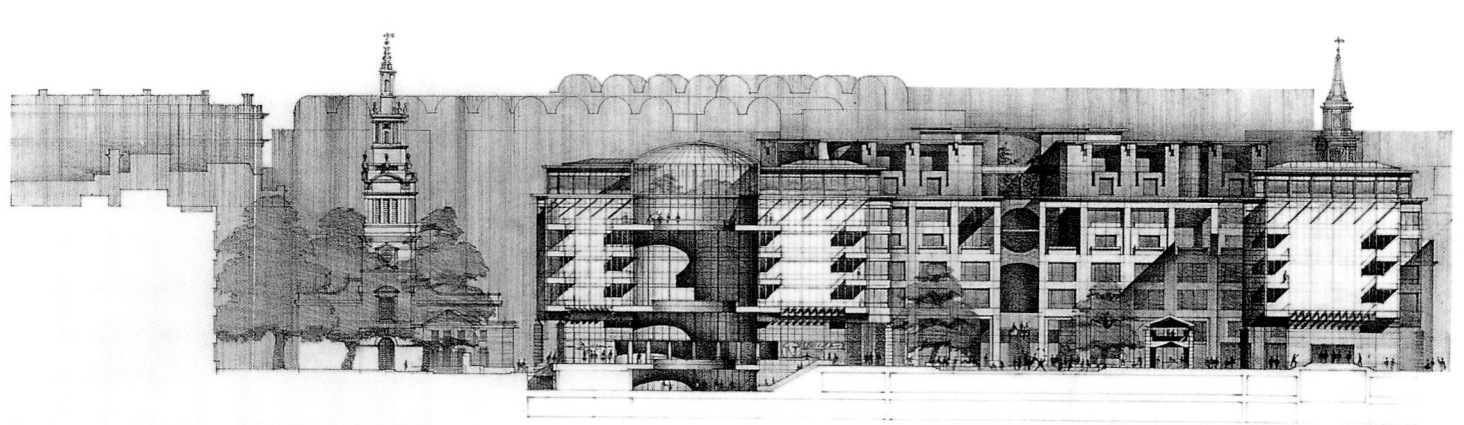

7

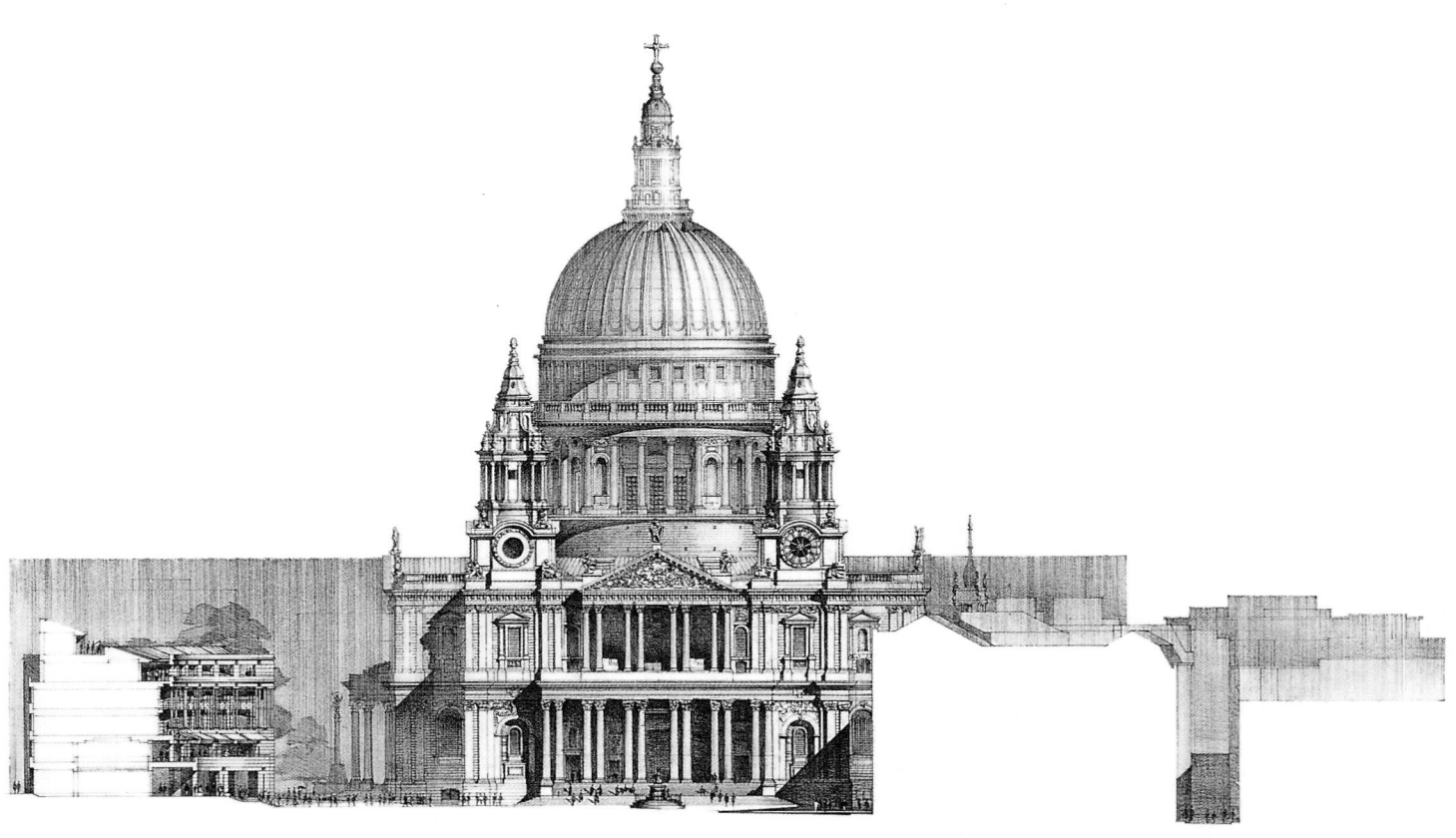

8

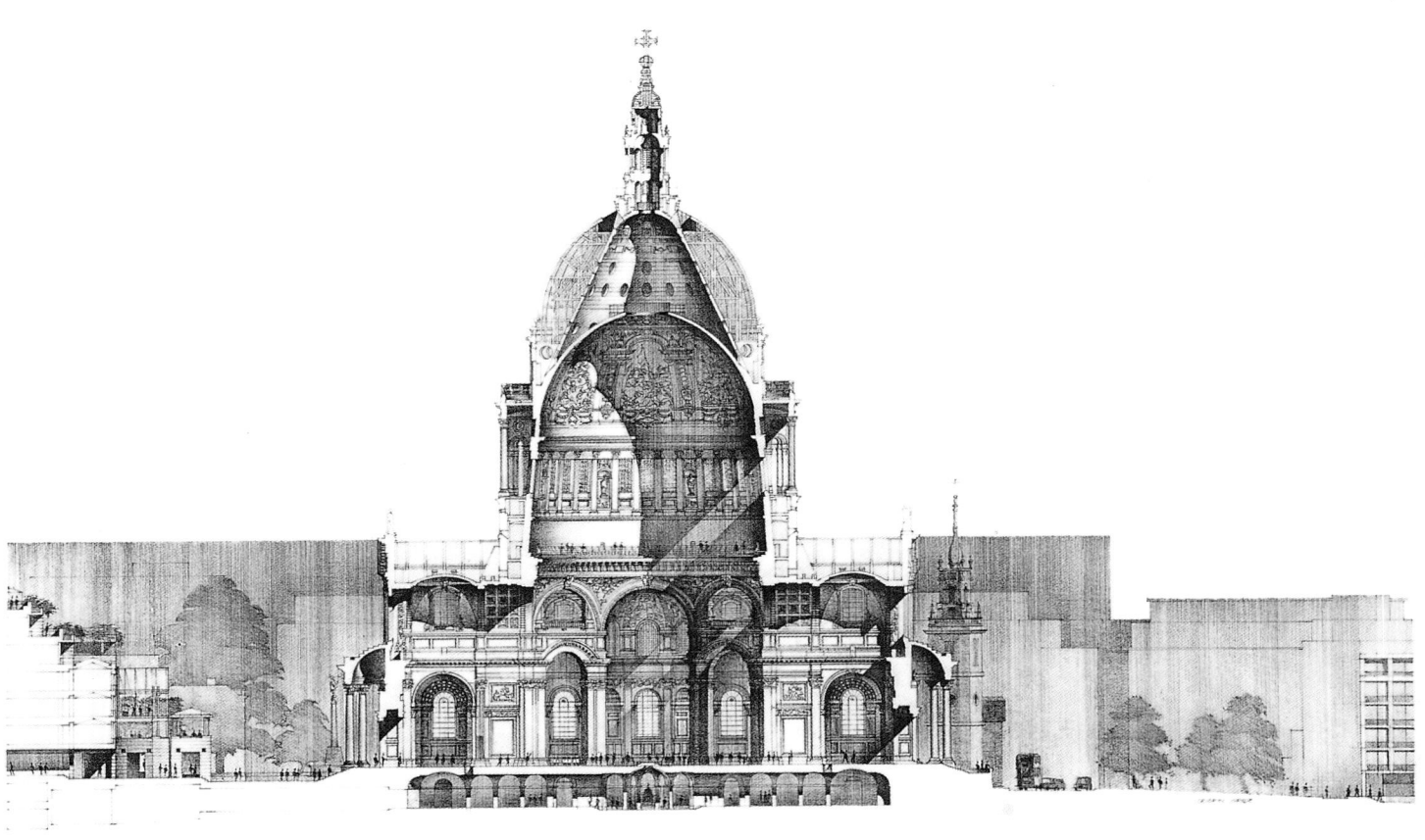

9

Paternoster Square 131

Stockley Park

Design/Completion 1984/1986
Heathrow, Uxbridge, Middlesex
The Stockley Park Consortium Limited
200,000 square metres
Office buildings: steel frame with metal panels and glass cladding
Arena: Forticrete block, metal roof and timber windows

斯托克利园区

设计/完成　　1984/1986
希思罗，优克斯桥，米德尔塞克斯郡
斯托克利园财团有限公司
200,000m²
写字楼：钢框架，金属镶板，玻璃外装
会所：砌体，金属屋顶，木窗

Stockley Park is a new business community and park created on the outskirts of London. Extensive land reclamation and careful site planning were required to establish a 50-hectare business park accommodating high-technology industries and a 100-hectare public park which includes an 18-hole golf course on the landfill site. The development was an important partnership between the public and private sectors.

Gravel extraction and tipping had been carried out on the site since before World War II. In 1981 an extensive site investigation was carried out by Ove Arup & Partners to determine the extent of the landfill and associated pollution.

The site planning strategy was to locate the business park in the south and create a green zone in the north. This not only enhanced the site for public use, but also created a landscaped strip between the new development and surrounding housing. The landfill was transferred from the business park site to the zone in the north. In the business park, gravel was used to provide stable building sites and to avoid the necessity of piled foundations

Continued

斯托克利园区建于伦敦市郊，是新型的商业社区和公园区。开发过程中对土地进行了大规模改造和精心规划。商业园区占地50hm²，其中安置高科技工业；公共公园区占地100hm²，包括建造在填土区的一块18洞的高尔夫球场。园区开发的重点在于处理这两部分即公共区和私人区之间的协调关系。

自二战以来，人们一直在这块场地中进行砾石的倾泄和处理。1981年阿鲁普联合事务所对其进行了大规模的场地勘测，以确定填土区的范围和土壤污染程度。

规划方案将商业区设在南部，在北部则规划了大片的绿化区。这增强了场地的公共用途，还在新建园区和周边建筑之间营造了一条景观带。北部填土区的回填土来自于南部的商业区场地。在商业园区，砾石可用于构造稳定的建筑场地条件，从而无需采用桩基础，也不必采用受污染场地通常设置的通风地下室。

待续

1 View of Arena building built on the edge of the main lake at the park entrance
2 Swimming pool in the Arena, with views along a landscaped valley
3 Section through curtain wall of first phase building
4 Main landscaped valley which provides an outlook for the business park buildings
5 Typical plan of first phase building

1 建于园区入口湖畔的会所外景
2 会所内的游泳池，由此可沿着山谷眺望远景
3 一期工程建筑幕墙剖面
4 山谷进行了景观设计，在其中可以望到商业园区的建筑
5 一期工程建筑标准平面

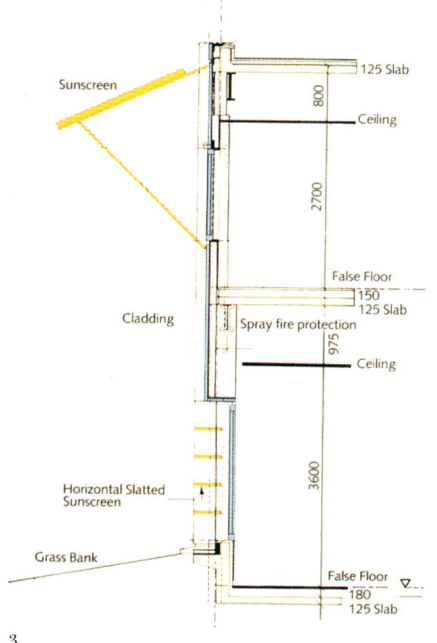

3

4

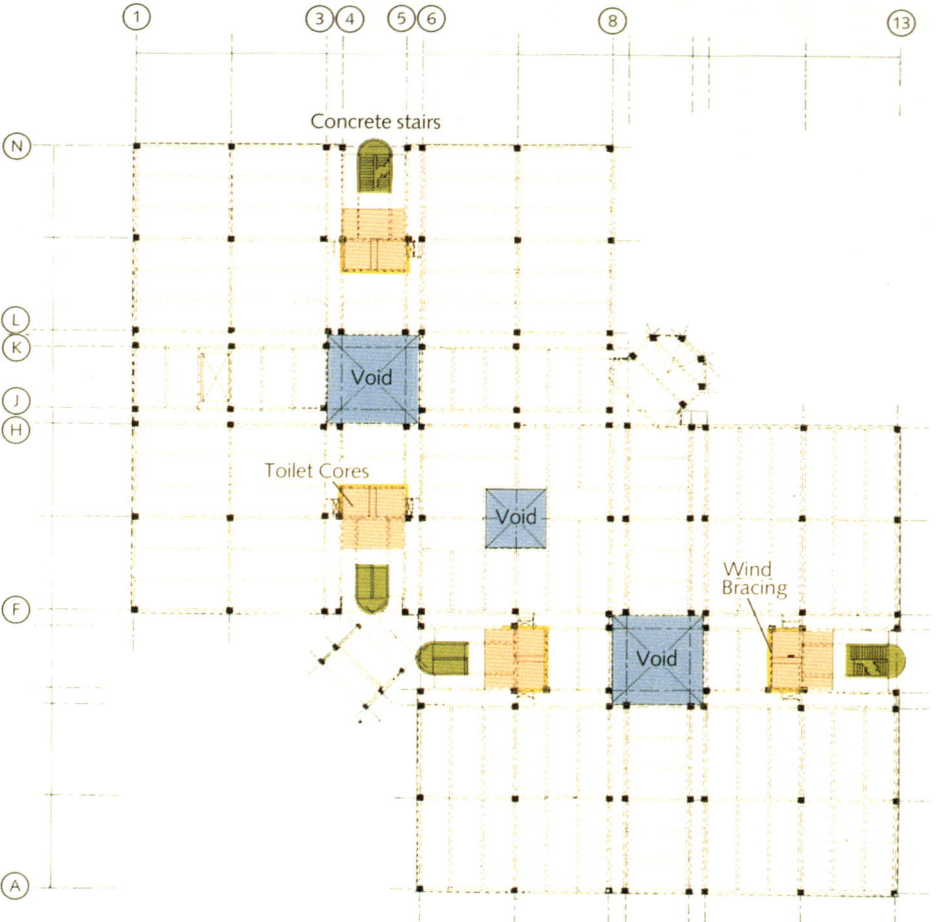

5

Stockley Park 133

and ventilated undercroft spaces normally required when building on polluted sites. A final layer of "manufactured topsoil" formed from sludge cake and clay was used to encourage growth in the newly planted landscapes. Approximately four million cubic metres of landfill, clay and gravel was moved to form new site contours for the business park and public open space zones.

The business park is made up of areas designated for new office and workshop buildings with a special area reserved for a central amenities building—the Arena (see page 156). The new buildings within Phase 1 of the business park are two- and three-storey structures providing a total of 140,000 square metres of gross floor area. A wide range of international companies including Apple, Tandem, Fujitsu, Toshiba, Dow, BP, Glaxo, Hasbro, EDS and Reebok occupy these buildings.

Construction started on the site in April 1985. The first three buildings were officially opened by HRH the Prince of Wales on 6 June 1986. Subsequently a further 17 buildings have been built at Stockley Park.

铺在新建的绿化景观区内表层种植土壤由污泥沉积物和黏土构成，可以促进植物的生长。约 400 万 m² 的回填土、黏土和砾石从商业区场地运送至公共开放园区堆填，构成了新的地面轮廓线。

商业园区由新写字楼和厂房建筑构成，中心地带还有一个特别的休闲娱乐场所——园区会所（见本书 156 页）。商业区一期工程的新建筑群一般为 2～3 层，可提供 140,000m² 的总建筑面积。众多的国际建筑公司参与了这些工程，包括苹果、坦德姆、富士通、东芝、道尔、BP、格拉斯科、哈斯伯罗、EDS 和锐步等。

工程于 1985 年 3 月动工。1986 年 6 月 6 日威尔士亲王亲自为第一批三幢建筑剪彩。随后在斯托克利园区又建造了 17 幢建筑。

6

6	Glazed atrium roof to restaurant in the Arena	6　会所中餐厅的玻璃天井顶棚
7	Landscaped walkway and bridge leading to a corporate office building	7　走道和小桥进行了景观设计，指引着去往公司写字楼的道路
8	Landscaped entry courtyard to the headquarters of Control Data	8　数控公司总部办公楼的入口庭院

7

8

Legal & General House

Design/Completion 1986/1988
Kingswood, Surrey
Legal & General Assurance Society Limited
24,000 square metres
Reinforced concrete frame
Metal and glazing infill panels, stone and brick cladding, timber screens

Designed to provide highly serviced and flexible offices, this new headquarters replaces an outdated office block which was built in the 1950s.

The three-storey building encloses two landscaped courtyards. Offices occupy the top two floors, with 8,000 square metres of ancillary support space including computer rooms, storage and restaurant facilities on the lower floor.

The project was designed with particular attention being given to energy efficiency. External sunscreens reduce solar gain, and the heat storage system is designed to utilise the staff swimming pool as a heat sink.

In addition to the new offices, St Monica's (a large Edwardian house on the site) has been renovated and extended to create a residential staff training centre for 46 visitors. The complex also includes recreational facilities with landscaped sports fields, tennis courts, sports pavilion, swimming pool and sports hall.

法人—通用保险公司大楼

设计/完成　　　1986/1988
金斯伍德，萨里郡
法人—通用保险公司
24,000m²
钢筋混凝土框架结构
金属和玻璃填充镶板，石块和砖块面层
木制帷幕

该项目要求总部大楼配有高水平配套服务设施，办公空间设置灵活多样。新建筑将取代已经陈旧过时的20世纪50年代建造的办公大楼供人们使用。

这幢三层的建筑物内含两个精心设计的庭院。办公室设在最上面的两层，另外设有8,000m²的附属设施如计算机房、贮藏室和底层的餐厅等。

工程的设计特别考虑了节能要求。外部帷幕减少了日照的影响，设置了蓄热系统，利用员工游泳池作为蓄热设施。

在新的办公楼以外，圣莫尼卡楼（场地原有的一幢爱德华时代建筑）也进行了重新修葺，并扩展成为可容纳46人住宿的员工训练中心。另外还建造了休闲娱乐场所，包括运动场、网球场、运动休息室、游泳池和运动馆。

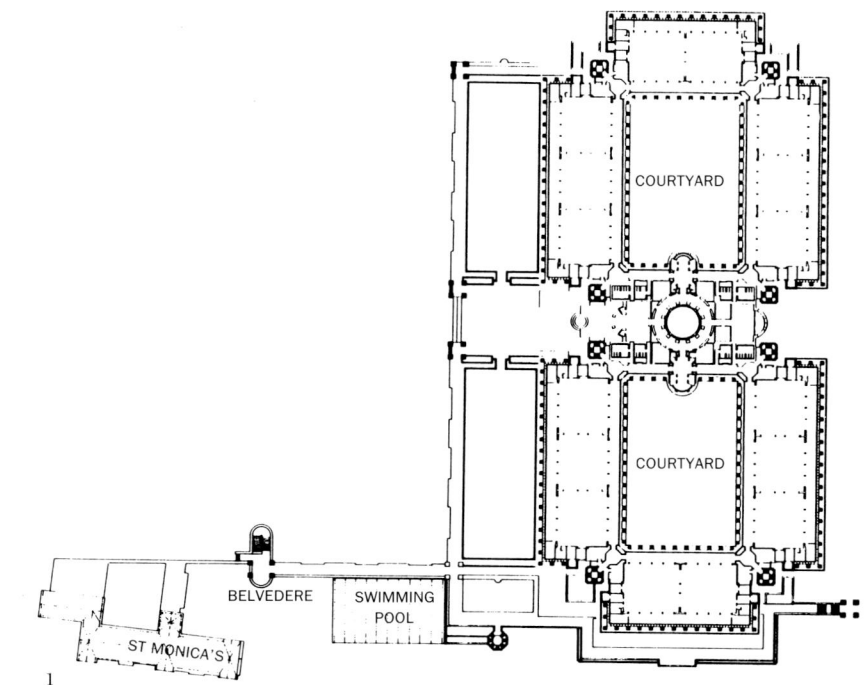

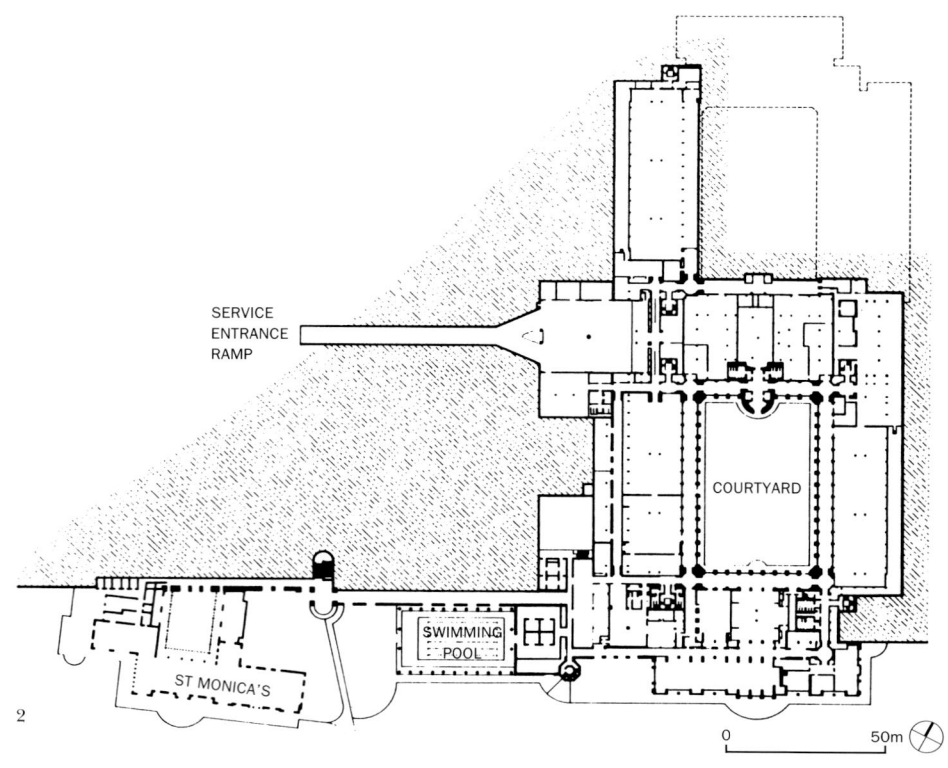

1 入口层平面图	1 Entrance level plan
2 底层/地下室平面图	2 Lower level/basement plan
3 上部办公室楼层平面图	3 Upper level office floor plan
4 从总部办公楼中往外眺望，可以看到新建的运动场地、网球场和运动休息室	4 The headquarters look out over new sports fields, tennis courts and a sports pavilion
5 办公楼主入口	5 The offices frame the main entrance

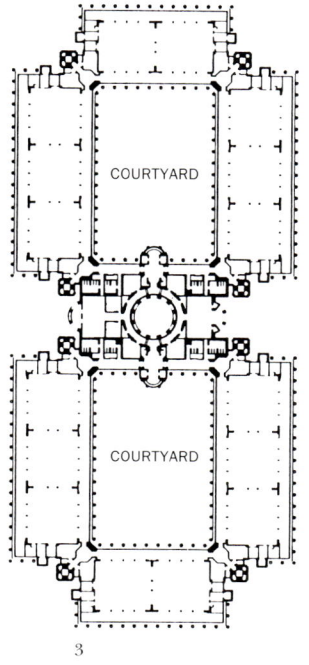

6 St Monica's, an exisiting Edwardian house, has been restored and adapted as a residential training centre
7 The belvedere on the upper terrace

6 场地原有的一幢爱德华时代建筑——圣莫尼卡楼，经过修葺后用作住宿式训练中心
7 位于阶地上部的观景亭

7

Legal & General House 139

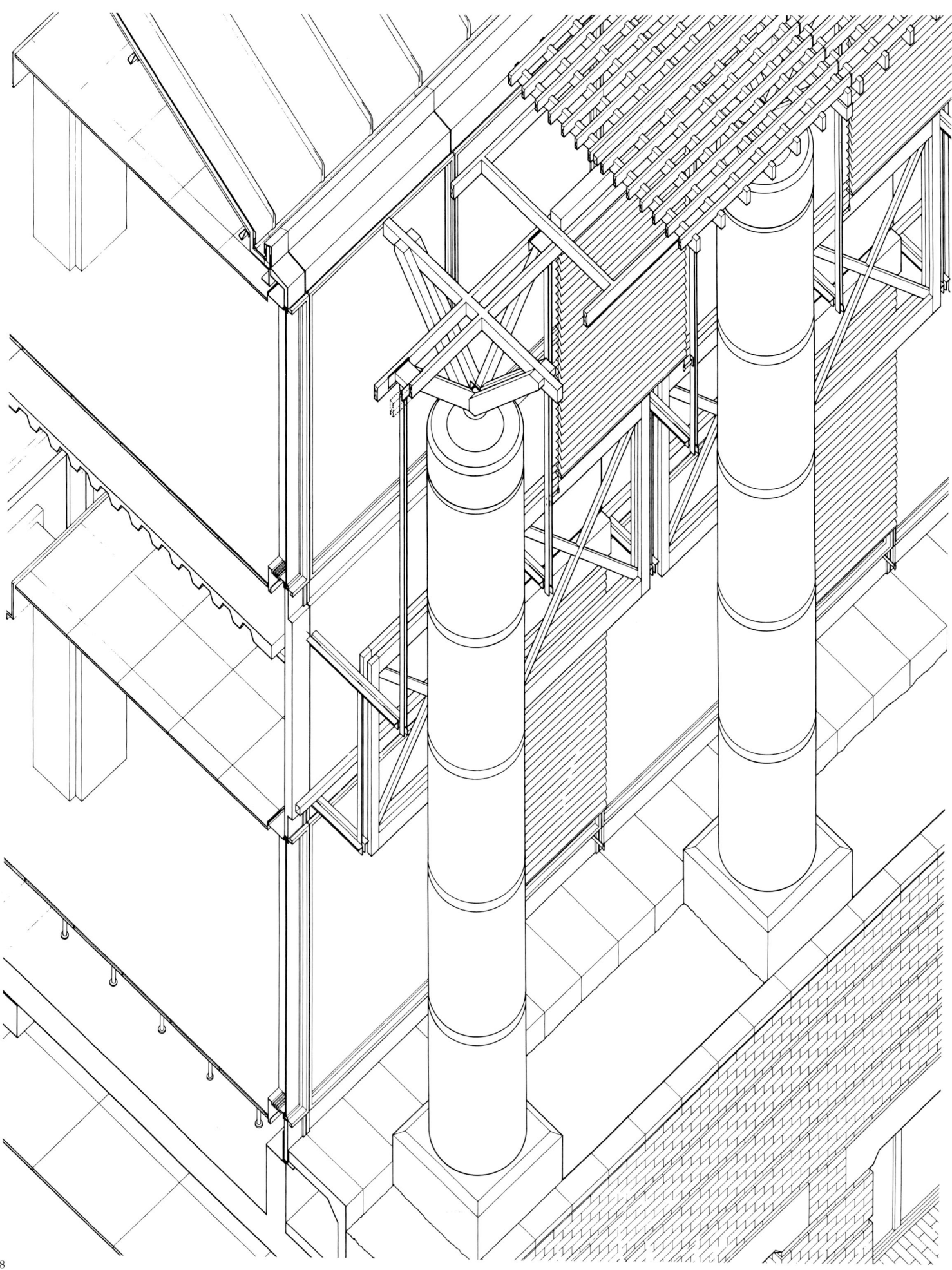
8

8 Dissected axonometric of outer wall and floors
9 The elevation creates a pergola with sunscreens and planting which protect and shade the glazed offices

8 外墙和楼板三向投影剖面图
9 外立面利用遮阳帷幕和植物搭成藤架，可以遮蔽阳光，保护玻璃幕墙后的办公室

10

10 员工游泳池
11 入口为圆顶石砌门厅,采用顶部采光

10 The staff swimming pool
11 A top-lit stone rotunda covers the entrance

12　Axial view through the rotunda
13　Inside the rotunda

12　圆顶门厅轴线视图
13　圆顶门厅内景

12

Hasbro Bradley (UK) Limited Headquarters

Design/Completion 1987/1988
Stockley Park, Heathrow, Uxbridge, Middlesex
Hasbro Bradley (UK) Limited
9,514 square metres
Steel frame
Metal cladding panels, glazing

哈斯伯罗—布拉德利（英国）有限公司总部

设计/完成　1987/1988
斯托克利园区，希思罗，优克斯桥，米德尔塞克斯郡
哈斯伯罗—布拉德利（英国）有限公司
9,514m²
钢框架结构
金属包层镶板，玻璃

The new European headquarters for the American company Hasbro Bradley (UK) Limited, one of the world's leading toy makers, is located on a focal site at the centre of Stockley Park in London. This building provides more than 9,500 square metres of space on two floors, focused around a glazed pavilion.

As well as different types of office work spaces, the headquarters accommodates a range of showrooms and a presentation suite specially designed for product demonstrations. An octagonal conservatory houses a coffee lounge with a mezzanine dining area seating 100.

The plan of the building has been organised around two linked blocks. These blocks consist of four pavilions, each 18 metres square, positioned symmetrically around two-storey high top-lit atria. The area provides a gallery for the client's extensive collection of paintings and sculpture.

The ground floor houses product engineering, including a workshop and facilities for research and development.

1

2

3

该项目为美国哈斯伯罗—布拉德利有限公司欧洲总部大楼，该公司为世界领先的玩具制造商。项目位于伦敦斯托克利园区的核心地段。大楼上下两层，建筑平面以中部的玻璃阁楼为中心对称布置，总建筑面积9,500m²。

由于不同的工作需要，大楼内设置一系列陈列室和为产品示范专门设计的展示室。一间八角形的暖厅内设置了咖啡室，暖厅内一层与二层之间设有夹层楼面，可供100人就座用餐。

整个建筑物由两栋相连的楼区组成，每个楼区又由四座18m正方的尖顶小楼组成，四座小楼围绕着中央的天井对称布置，天井高度和建筑一致，采用顶部采光。天井四周设置了廊道，供业主展示美术和雕塑等艺术收藏品。

生产用房位于首层，包括一个车间和研发部门。

1　General view of the building with the octagonal restaurant overlooking a lake
2　Central atrium with display areas on the ground floor
3　Central atrium with first floor access bridge
4　View of lakeside facade
5　View from octagonal restaurant
6　Facade detail with sun control devices

1　建筑物外景，从八角形餐厅中可以看到楼外的湖水
2　中央天井，在首层设有展示区
3　中央天井，二层中间架有小桥作为走道
4　建筑临湖立面外景
5　从八角形餐厅看到的景色
6　立面细部，配有采光控制设施

Hasbro Bradley (UK) Limited Headquarters 147

The Imperial War Museum, Stage I

Design/Completion 1983/1989
Lambeth, London SE1
Property Services Agency
8,000 square metres
Steel roof
Concrete structural slabs, ceramic tiles

The Imperial War Museum is housed in the former Bethlem Royal Hospital designed by James Lewis and completed in 1815. It is listed as a Grade 2 Historic Building.

In 1983 Arup Associates were appointed to prepare a feasibility study for the redevelopment of the Imperial War Museum. The brief was to significantly increase the amount of gallery space, to meet the strict environmental conditions appropriate for a national museum holding a large and diverse collection, and to improve facilities for visitors.

The first stage of the redevelopment created 4,600 square metres of new floorspace on four levels around a new central, top-lit exhibition hall which forms the focus of the new Imperial War Museum. A light steel structure supports a diagonal lattice barrel vault designed to allow aeroplanes and other large exhibits to be suspended within. This delicate white-painted structure has been built within the old courtyard.

Continued

帝国战争博物馆建于原来的贝斯勒姆皇家医院内，这幢被列为二级保护单位的建筑物1815年建成，设计者为詹姆斯·刘易斯。

1983年阿鲁普联合事务所接受委托，对帝国战争博物馆的改造扩建进行可行性分析。改造规划重点在于增大陈列廊道的空间，以适应日益增长和种类繁多的博物馆藏品的陈列需要，另外还计划改进游客服务设施。

改造工程的第一阶段包括建造新的中央展厅，展厅采用顶部采光，将成为新博物馆的中心。展厅四周将营造上下4层共4,600m² 的新楼面空间，采用轻钢结构建造一个交叉格构式拱顶，可以用于悬挂飞机和其他大型展品。这个精巧的白色拱顶结构建造在原来的庭院中，原来的医院建筑也进行了粉刷和修复。

帝国战争博物馆一期工程

设计/完成　　　1983/1989
朗伯斯区，伦敦西南一区
公共资产机构
8,000m²
钢屋盖
混凝土结构板，陶瓷砖瓦

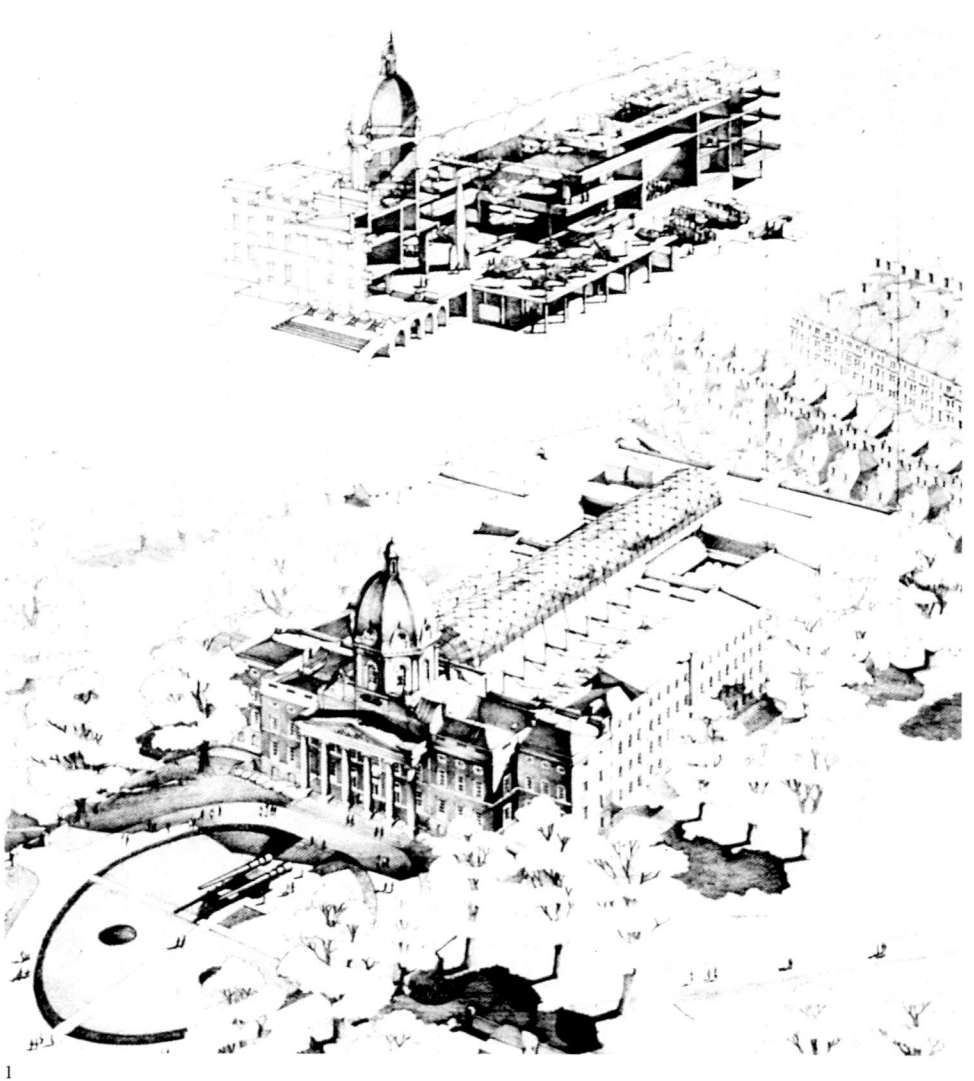

1

1 The master plan for the development of the museum creates a new sky-lit hall built in the courtyard of the restored historic building
2 The new vaulted steel structure contrasts with the restored historic building

1 博物馆改造工程规划在原有古建筑的庭院中建造一座新的中央展厅，展厅采用自然采光
2 新型的钢结构拱顶与修复后的古建筑形成对比效果

The Imperial War Museum, Stage I   149

3　The central hall has been designed for the display of large exhibits
4　Lower ground-floor plan
5　Ground-floor plan
6　First-floor plan

3　中央展厅设计用于展出大型展品
4　地下层平面图
5　底层平面图
6　二层平面图

The former hospital building has been cleaned and restored.

As well as spaces for large exhibits (including aircraft, rockets, submarines and artillery) there is a total of 1,500 square metres of fully serviced display space. At the lower ground floor level and on the upper floors, there are two new suites of galleries designed to accommodate special exhibitions. These include the museum's unique collection of paintings, drawings and sculpture—a collection which includes works by Stanley Spencer, Paul Nash and Henry Moore and which is the second largest collection of twentieth century British art in the country. In addition there is a new main entrance hall, a licensed cafe and an extensive shop.

The first stage in the museum's major redevelopment was opened by H. M. the Queen on 29 June 1989. The next stage consists of four floors of new galleries with both artificially lit and day-lit spaces. It will be completed in 1995.

在能够为大型展品（包括飞机、火箭、潜水艇和大炮）提供展览空间的同时，建筑还包括总共1500m²的有完备配套设施的展示区。在地面层和上部楼层，建造了两套新的展示廊道可以用于举行特别的展览，包括博物馆特有的书画和雕塑等藏品——博物馆拥有英国第二的20世纪英国艺术品馆藏，包括斯坦利·斯宾塞、保罗·纳什和亨利·摩尔等人的作品。另外，建筑还包括新的主入口门厅、一间专营咖啡馆和一间商店。

1989年6月29日，博物馆主改造工程一期工程在女王陛下的主持下剪彩并向公众开放。二期工程将包括4层新的陈列廊道，同时采用人工照明和自然采光，将于1995年完工。

3

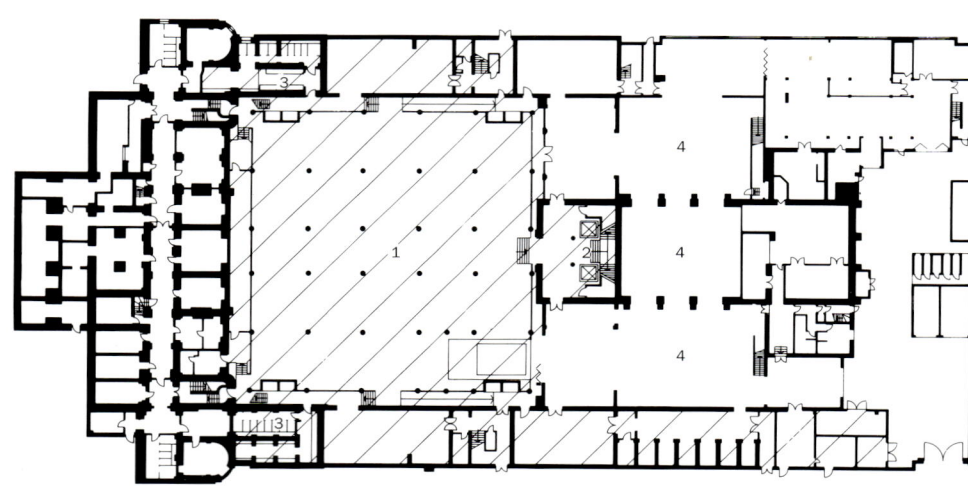

4

　　Redevelopment
1　Exhibition space
2　Main stairs/lifts
3　Public lavatories
4　Exhibition galleries

　　改造工程
1　展览区
2　主要楼梯/电梯
3　公用洗手间
4　陈列廊道

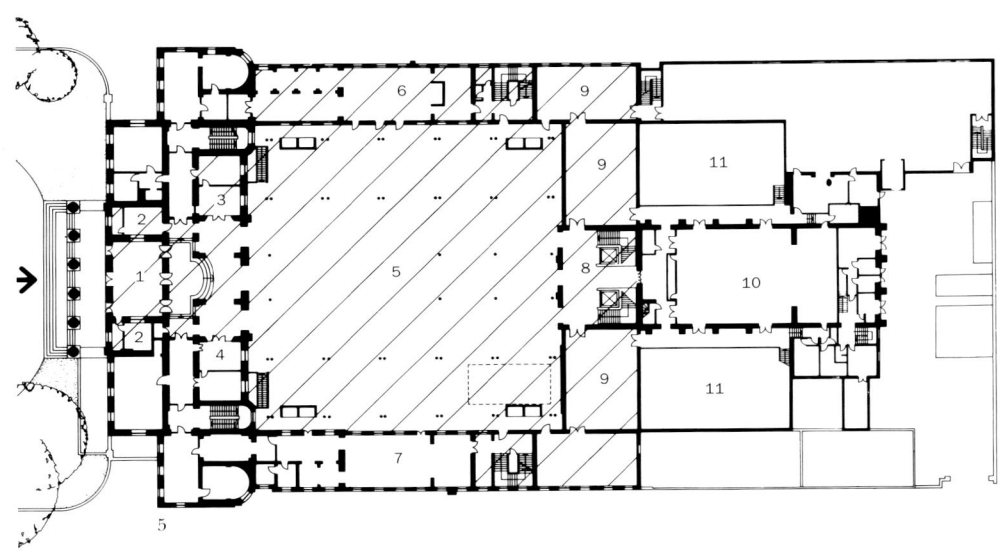

╱╱ Redevelopment	6 Cafe	改造工程	6 咖啡室
1 Entrance	7 Shop	1 入口	7 商店
2 Cloakrooms	8 Main stairs/lifts	2 卫生间	8 主要楼梯／电梯
3 Reception	9 Art galleries	3 接待区	9 艺廊
4 Information	10 Cinema	4 消息栏	10 放映厅
5 Main exhibition	11 Exhibition gallery	5 主展示区	11 陈列室

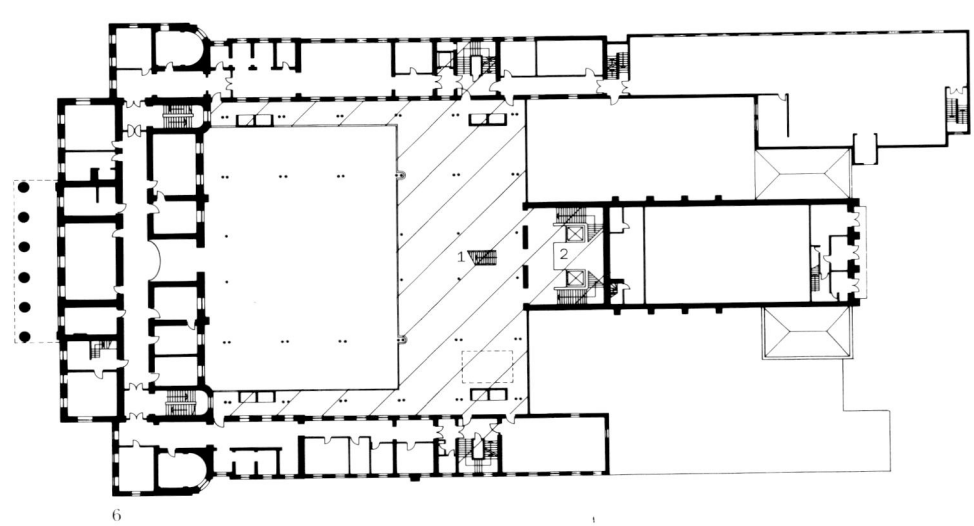

╱╱ Redevelopment		改造工程
1 Exhibition gallery		1 陈列廊道
2 Main stairs/lifts		2 主要楼梯／电梯

The Imperial War Museum, Stage I   151

7　Cross section
8　Long section
9　The new front entrance

7　横截面图
8　纵截面图
9　新正门入口

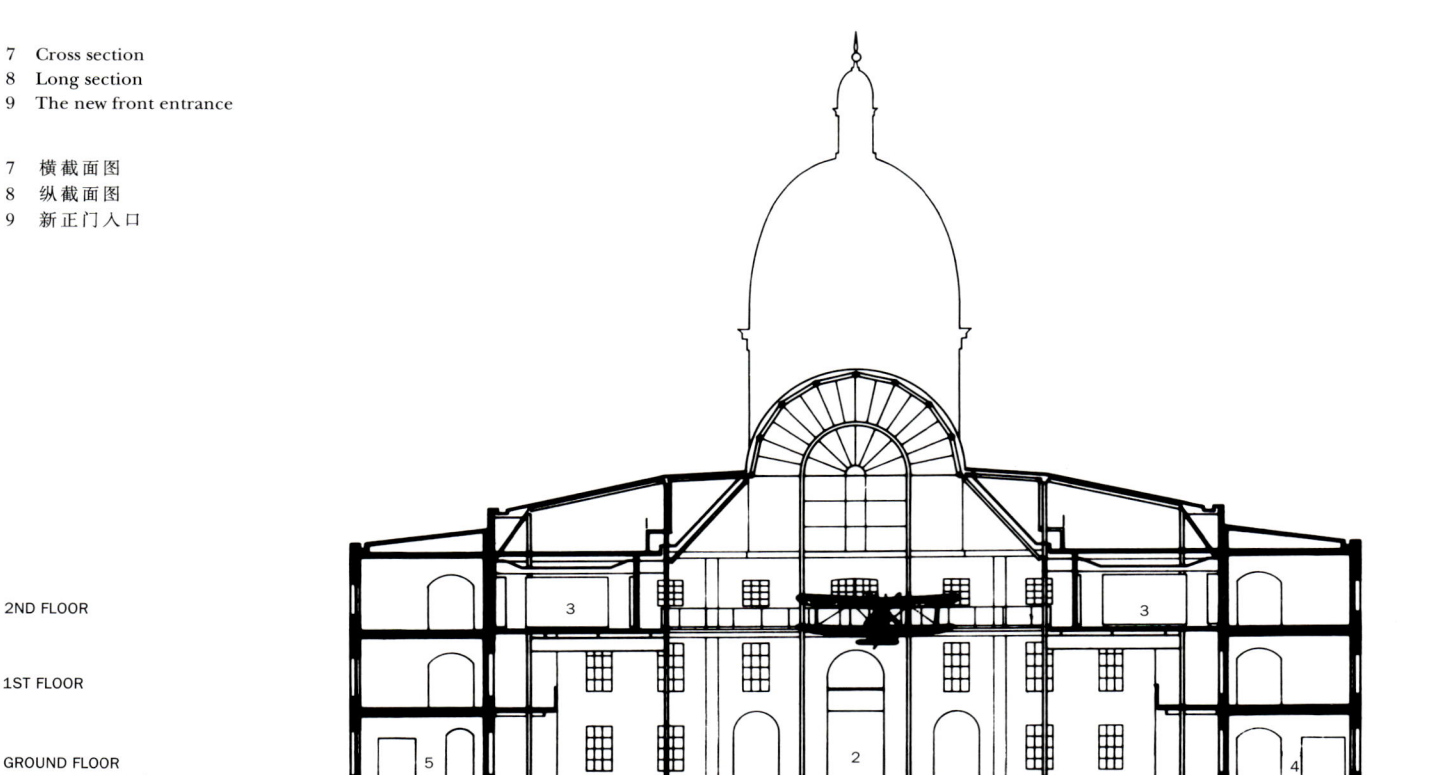

2ND FLOOR
1ST FLOOR
GROUND FLOOR
LOWER GROUND FLOOR

7

1　Lower ground floor-permanent exhibition
2　Ground floor-main exhibition space
3　Second floor-art galleries
4　Cafe
5　Shop

1　地下室，展出固定不变展品
2　一层——主展区
3　三层——艺术品展区
4　咖啡室
5　商店

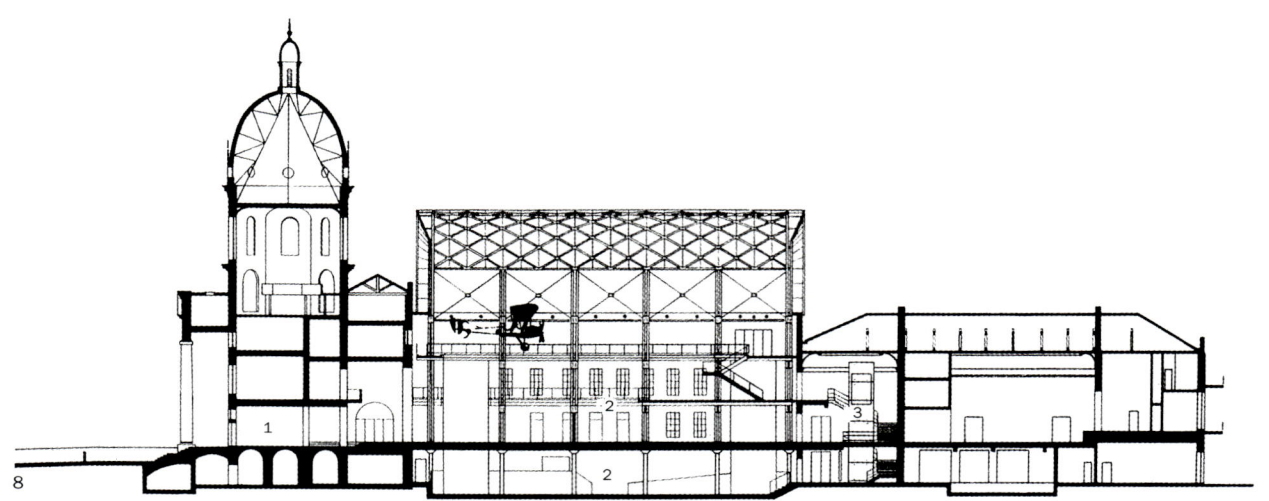

8

1　Main entrance
2　Exhibition spaces
3　Main stairs/lifts

1　主入口
2　展览区
3　主要楼梯/电梯

9

10

10 Within the central hall the design creates a series of stepped galleries
11 The day-lit hall is the focus of the new museum
12 The steel structure has been clearly expressed
13 The new entrance hall contained within the restored historic shell
14 The new cafeteria located within the existing building
15 New art galleries overlook the central hall

10 在中央大厅四周设计了系列的多层陈列廊道
11 大厅采用自然采光，是新博物馆的中心
12 钢结构直接的展露在外
13 壳形古建筑改造成新的入口门厅
14 新咖啡厅位于原有建筑中
15 新的艺廊，由此可以俯瞰中央大厅的景象

11

12

13

14

15

The Imperial War Museum, Stage I 155

The Arena, Stockley Park

Design/Completion 1988/1989
Heathrow, Uxbridge, Middlesex
The Stockley Park Consortium Limited
6,000 square metres
Precast masonry block (Fiagreca)
Masonry walls, hardwood windows, lead-coated stainless steel roof

斯托克利园区会所

设计/完成　　1988/1989
希思罗，优克斯桥，米德尔塞克斯郡
斯托克利园财团有限公司
6,000m²
预制砌体
圬工墙体，硬木窗，包铅不锈钢屋盖

The Arena is the social hub of Stockley Park (see page 132). The two-storey building, located between the new golf course and the business park, houses a mix of sports and social facilities, shops, a restaurant and wine bar, together with the management offices.

Located alongside a newly created lake, the building has been constructed in the first phase of development to provide a wide range of amenities for a working population of some 4,500 people. The facilities are also available for the use of the local community.

The building has been organised around a central circular colonnaded courtyard. The sports facilities—a sports hall, a gymnasium, four squash courts and exercise rooms—have been planned on the ground floor, whilst shops, including a bank, travel agent, dry cleaners and a newsagent, front onto the central courtyard. The windows of the restaurant look out over the lake. On the upper floor are conference facilities, the offices of the management and a sports club bar which overlooks a swimming pool.

该项目为斯托克利园区（见本书132页）的社交活动会所，项目位于新高尔夫球场和商业园区之间。建筑上下两层，楼内包含运动和社交场所、商店、一家餐厅和酒吧，同时还有会所管理办公用房。

会所位于新开挖的人工湖畔，于园区一期工程建造。建成后为园区内约4,500人提供了多样的休闲娱乐服务。园区当地居民也可以享用其中的设施和服务。

建筑物环绕中央庭院而建，庭院带有柱廊。运动设施包括一间体育馆、一间健身房、四间壁球馆以及锻炼房，都设在第一层。商店、银行、旅行社、干洗店和一家书报屋，也同时面朝中央庭院而设。餐馆中可以凭窗眺望楼外的湖水。上部楼层中设有会议室、管理办公室和一间运动俱乐部酒吧，从二层可以俯瞰到游泳池。

1

2

1　餐厅里的酒吧和休闲室
2　游泳池及远处的景观
3　湖畔的人行小道

1　Bar/lounge in the main restaurant
2　Swimming pool with views of the landscape beyond
3　Pedestrian walkway along the lake edge

Selected and Current Works

作品精选

1990s

Sussex Grandstand, Goodwood

Design/Completion 1987/1990
Goodwood, Sussex
Goodwood Racecourse Limited
5,000 square metres
In situ coffered concrete slab; steelwork
Fabric roof; polyester reinforced PVC; brick base to first storey, glass

苏塞克斯正面看台，古德伍德赛马场

设计/完成　　1987/1990
古德伍德，苏塞克斯郡
古德伍德赛马场有限公司
5,000m²
现浇混凝土板，钢结构
膜屋顶，聚酯加劲PVC，砖基础（至一层），玻璃

The Sussex Grandstand at Goodwood was opened in 1990. This new stand marks the completion of the second phase of a master plan prepared by Arup Associates in 1987 for the development of this well-known horse racing course. The grandstand was designed to provide space for more than 3,000 spectators in both terraces and seating areas. Following detailed studies of the client's requirements, the design of this scheme incorporates a light fabric canopy which gives protection from wind and showers as well as providing shade on sunny summer days.

The new grandstand, situated on a prominent site on the Downs, provides elevated seating terraces under a lightweight roof. A white steel structure supports the upper level seating areas. The boxes and sponsors' accommodation are at first floor level. The grandstand sits on a stepped brick mound which provides additional viewing areas for spectators.

Refreshment and catering facilities for spectators are also included within the Sussex Grandstand, as well as private hospitality suites, tote and off-course betting facilities, cloakrooms and administrative offices.

The lightness of the structure and its canopy captures the spirit of festivity which is so typical of summertime events such as Glorious Goodwood.

古德伍德赛马场的苏塞克斯正面看台于1990年落成，这一看台的建成标志着这个著名赛马场的改造二期工程的完成。1987年阿鲁普联合事务所就对这项工程进行了规划，看台可以容纳3000多名观众入座观看比赛。在仔细研究了用户需求后，看台的设计采用了轻质的膜顶棚，可以使观众免受夏天的日晒和风雨之苦。

新的看台建造在一片开阔的高地上，看台为台阶形，采用轻质屋顶。上层的座位区采用白色的钢结构承重。包厢和嘉宾席设在首层。看台下方的土丘坡上用砖砌建造了观众席，也可供观看比赛使用。

看台中也配备了为观众提供服务的餐饮设施、私人休息室、赌马投注站、卫生间和管理办公室等。

看台屋盖轻快的结构形式符合了赛马比赛激烈欢快的气氛，每年夏天这里都会举行盛大的古德伍德赛马会。

1

2

3

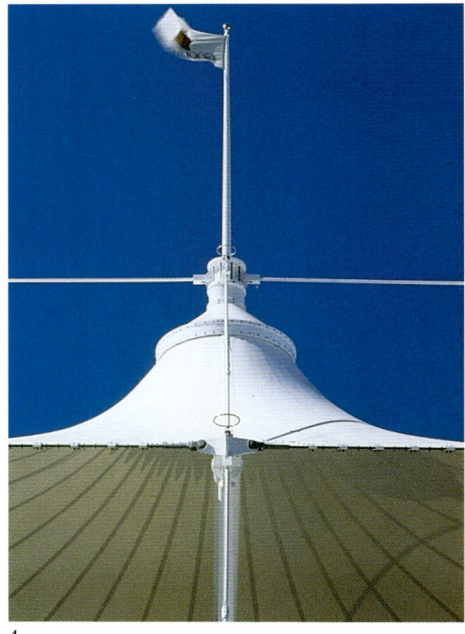

4

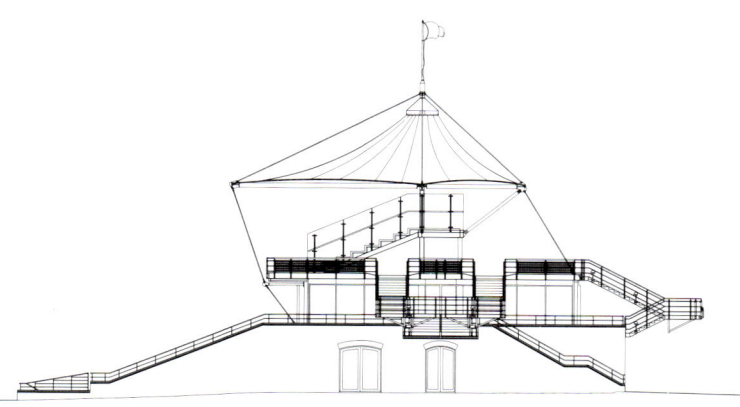
5

6

8

7

1 View of the end of the terrace roof
2 Race day
3 Detail of fabric roof
4 Roof structure detail
5 Side elevation through grandstand
6 Model of the grandstand
7 View looking out at course from hospitality suite
8 View of the grandstand in operation from bookmakers' forecourt

1 看台屋顶端部视图
2 赛马大会
3 膜屋顶细部
4 屋顶结构细部
5 看台侧立面
6 看台模型
7 从休息室内俯瞰马场跑道
8 使用中的看台外景

Sussex Grandstand, Goodwood   161

Wentworth Golf Club

Design/Completion 1989/1990
Wentworth, Surrey
Chelsfield PLC/Wentworth Holdings
8,500 square metres
Steel, brick and timber
Slate roof

温特沃斯高尔夫球俱乐部

设计/完成　　1989/1990
温特沃斯，萨里郡
切尔西菲尔德公司/温特沃斯资产机构
8,500m²
钢结构，砖，木材
石板屋顶

Following their successful submission in an invited competition involving six international architectural practices in March 1989, Arup Associates were asked to develop detailed design proposals for the new clubhouse and sporting facilities at Wentworth Golf Club in Surrey.

The proposals were based on a master plan for the site which established locations for a new 7,400 square metre golf clubhouse, screened parking and improved facilities for the tennis club, within a landscaped framework.
The golf clubhouse was designed on three levels, providing restaurant facilities, a health and fitness spa, together with meeting rooms, club offices, lockers and a shop. The restaurant and lounges were planned to overlook the existing green and fine woodlands on the site, whilst the overall form of the building was kept low to respect its natural setting.

1989年3月阿鲁普联合事务所在6家国际建筑公司参与的邀请设计竞赛中脱颖而出，承担了萨里郡温特沃斯高尔夫球俱乐部的新建俱乐部建筑和运动场项目的设计规划。

设计方案对场地平面进行了规划，计划建设占地7,400m²的高尔夫球俱乐部会所、有遮蔽的停车坪和改进的网球馆设施，整个区域将进行景观设计和绿化。高尔夫球会所设计为3层，内含餐厅、健身房和温泉浴室、会议室、俱乐部办公室、更衣室和商店。餐厅和休息室中可以眺望场地原有的树林，建筑体形将限制高度以与周边自然环境相协调。

待　续

1

1　The golf clubhouse has views over the course to the lake beyond
2　The new tennis club is an important element in the master plan for the Wentworth Golf Club

1　高尔夫球俱乐部，穿过球场可眺望远处的湖水
2　新建的网球俱乐部是温特沃斯高尔夫球俱乐部总规划中的重要部分

The new tennis pavilion is sited alongside a group of 14 tennis courts. It provides a range of facilities including a lounge and bar with locker rooms, tennis professional's office and a small shop for sporting equipment. The single-storey building has been planned to overlook the five new courts. The nine existing courts have been refurbished.

An outdoor terrace, partly sheltered by the overhanging roof of the pavilion, provides space for both players and spectators, with good views of the playing areas. Framed in Douglas Fir with double beams of laminated redwood supporting a pitched roof of slate, the scheme attempts to develop and extend the tradition of timber pavilions. The scheme creates a series of spaces which are finished in white and natural timber. The steel roof structure is exposed and contrasts with the ceiling of Douglas Fir. The floor is finished in beech.
Arup Associates were responsible for both the architectural and interior design.

新的网球馆位于俱乐部旁边，共有14块网球场。它也配备了一系列配套设施，如休息室、酒吧、更衣室、网球教练办公室和一间体育装备小店。单层的网球馆中可以看到室外的5块新网球场地，原有的9块网球场也进行了整修。

另外，还建造了一座户外的露台，露台部分采用悬挑式的屋顶遮挡，可供比赛者和观众更好的观看比赛。设计试图拓展传统木结构的使用，石板坡屋盖的承重体系采用了道格拉斯杉木和红木层压板组成的框架结构。室内设计大量采用了白色天然木材进行装饰。钢屋盖结构外露，与道格拉斯杉木顶棚形成了鲜明对比。地板采用山毛榉木。阿鲁普联合事务所同时负责了该项目的建筑设计和室内设计。

3

4

3 Sectional model showing the design for the new golf clubhouse
4 The tennis pavilion and grass courts
5 The approach to the tennis pavilion
6 Interior of the pavilion showing bar area

3 高尔夫球俱乐部会所建筑断面模型
4 网球馆和草地球场
5 去往网球馆的通道
6 网球馆内部酒吧内景

5

6

Arena, Bergiselstadion

Competition 1991
Bergisel, near Innsbruck, Austria
Municipality of Innsbruck
20,000 seating capacity
Reinforced concrete and steel
Glazing and lightweight roof panels

伯格塞尔露天赛场

设计竞赛时间　　1991
伯格塞尔，因斯布鲁克附近地区，奥地利
20,000 座位
钢筋混凝土和钢结构
玻璃，轻质屋顶面板

The design for a new arena at the ski jump in Bergisel, near Innsbruck in Austria, includes a retractable cover for the arena together with improved facilities for competitors, spectators and press.

A ring is created around the rim of the existing Bergisel stadium, unifying its geometry and providing a promenade for visitors to enjoy the magnificent views. Twelve columns support the ring, each one containing a staircase providing access from the bars and washrooms below.

An inclined conical mast, the "torch", extends from the ring to support the roof. Visitors can take an elevator up the mast to an observation loft where the Olympic flame burns. The roof consists of 24 wing-like vanes that can be extended to provide cover ranging from 90 to 360 degrees.

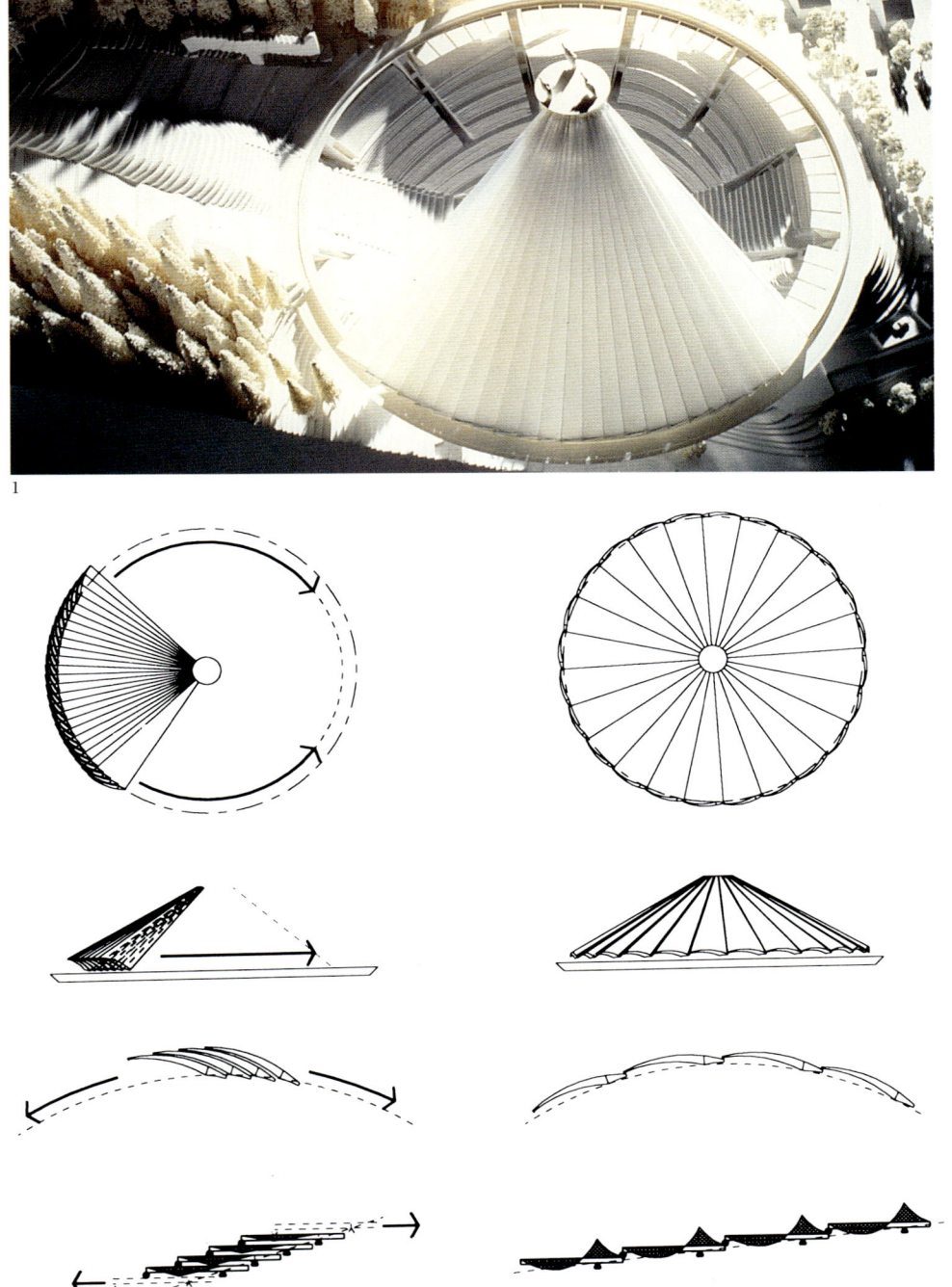

1

　　这个赛场是为奥地利因斯布鲁克附近地区举行的跳台滑雪比赛建造的。工程包括可收缩的赛场天篷，以及为运动员、观众和新闻记者配备的配套设施。
　　建筑的主要构成部分是一座环绕原有的伯格塞尔运动场的边缘建造的圆环，这使新旧场地在几何外形上成为一体。游客还可以沿着环形的走道散步观赏壮观的景色。12根柱共同支承起这个圆环，每根柱旁建造了上下楼梯，通往圆环下方的酒吧和卫生间等地。
　　从圆环向外伸出一根倾斜的圆锥形桅杆，支撑起屋盖。游客可以通过电梯上升到桅杆处，观看置于阁楼中的燃烧着的奥林匹克火炬。屋盖由24片翼状的叶片构成，能够进行收缩，覆盖90°～360°范围内的区域。

2

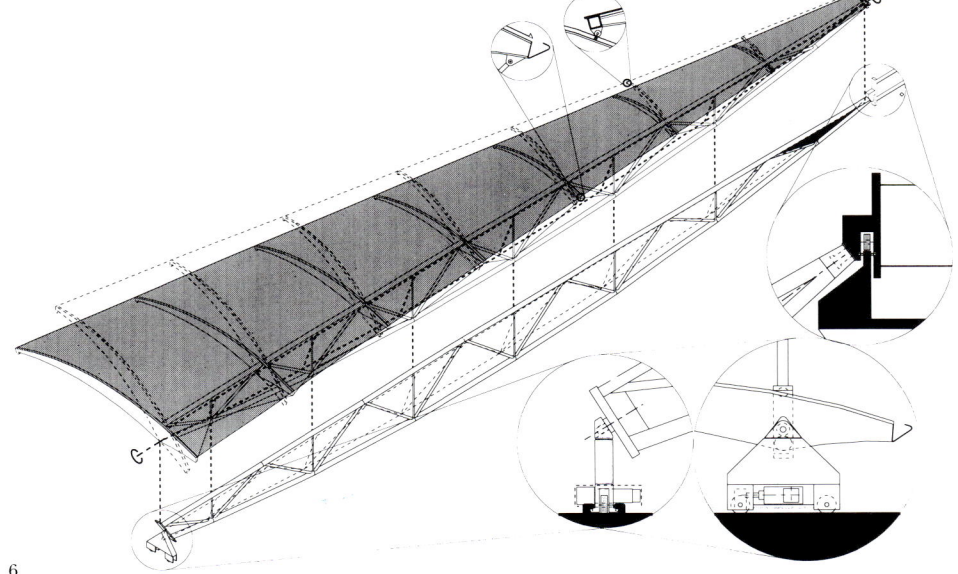

1	Aerial view of the model with ski slope to the left	1	模型鸟瞰图，左为滑雪斜坡
2	Anatomy of roof mechanism	2	屋盖机构解剖图
3	Cross section showing profile of ski jump and proximity of the bowl	3	高台滑雪场及碗状赛场剖面图
4–5	Model showing roof in "parked" position	4–5	屋盖"收回"状态模型
6	Anatomy of roof mechanism	6	屋盖机构解剖图

Arena, Bergiselstadion

Offices for Lloyds Bank PLC

Design/Completion 1988/1992
Canons Marsh, Bristol
Lloyds Bank PLC
33,500 square metres
Concrete frame and coffered slab
Glass roof on steel columns, stone, precast concrete, iroko, glazed curtain wall; lead-coated stainless steel

劳埃德银行股份公司办公大楼

设计/完成　1988/1992
卡农斯马什，布里斯托尔
劳埃德银行股份公司
33,500m²
钢筋混凝土框架结构，镶板
玻璃屋盖，钢柱，石料，预制混凝土
玻璃幕墙，包铅不锈钢

The new headquarters for Lloyds Bank Retail Banking in Bristol was designed both to provide offices and to create a new civic space in the city.

The 4.5-hectare site within Bristol's docklands was formerly occupied by a group of large bonded warehouses which overshadowed the area and cut off views of the cathedral and the city. The warehouses were demolished in 1988.

The new, three-storey development has been carefully designed to frame a series of views of the cathedral and the harbour.

The building provides work spaces for some 1,400 staff, as well as computer rooms, an entrance hall, a restaurant and underground car parking. Externally the scheme also includes extensive terraces and paved public amenity areas with landscaped parking.

Continued

该项目为劳埃德银行（小额存放业务）公司在布里斯托尔的新总部大楼，大楼作为办公室的同时，也为市民提供公共活动空间。

场地占地4.5hm²，位于布里斯托尔港口区，过去这里建造了大片的保税品仓库，影响了附近的大教堂和整个城市的景观。1988年仓库全部被拆除。

新建的3层建筑进行了周密的设计，试图依托教堂和港口营造一系列景观。

大楼内可容纳1,400名员工在其中办公，同时也包含计算机房、入口门厅、一间餐馆和地下停车场。在大楼外，建造了宽阔的景观露台、草坪和公共休闲区域。

1

1　卡农斯马什鸟瞰图
2　海港
3　教堂新景，位于一期工程和二期工程之间

1　Aerial view of Canon's Marsh
2　The floating harbour
3　New view of the cathedral between phase 1 and phase 2

Offices for Lloyds Bank PLC 169

The design uses stone, concrete and timber—materials which have been selected and finely crafted to create a civic presence. The external wall also incorporates screening with maintenance walkways to the large glazed areas, reflecting concerns for energy conservation and efficiency in use.

Other aspects of the design which have been developed to conserve energy include the development of compact building forms; the use of exposed concrete ceilings which maximise the thermal capacity of the structure; the introduction of daylight into the heart of the building which reduces the demand for artificial lighting; floor air supply with high-level extraction which maximises natural convection; and the use of dock water for cooling.

建筑材料选用了石料、混凝土和木材，这些材料进行了精心筛选和处理，以适合公共空间的要求。外墙采用玻璃幕墙，反映出节能和使用效率方面的考虑。

大楼的设计还在其他方面考虑了节能，包括采用紧凑简洁的建筑形式，采用外露的混凝土顶棚增强混凝土的蓄热能力；另外，设计使建筑内部可得到自然光线而减少了人工照明的需要，采用了高效的地板通风系统加强空气的自然对流，直接利用港口的水源制冷等等。

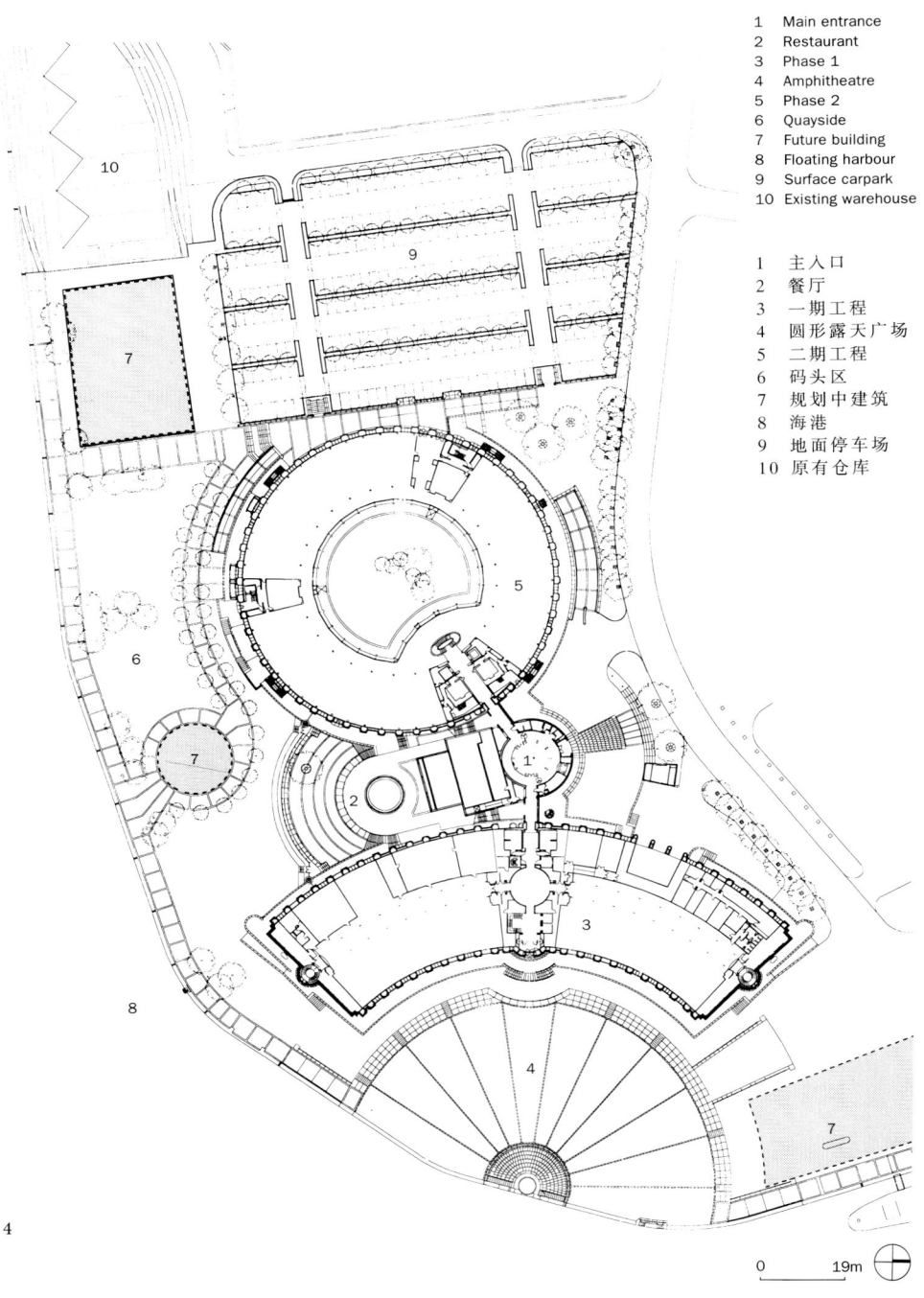

1 Main entrance
2 Restaurant
3 Phase 1
4 Amphitheatre
5 Phase 2
6 Quayside
7 Future building
8 Floating harbour
9 Surface carpark
10 Existing warehouse

1 主入口
2 餐厅
3 一期工程
4 圆形露天广场
5 二期工程
6 码头区
7 规划中建筑
8 海港
9 地面停车场
10 原有仓库

4 Site plan
5 View across the floating harbour towards phase 1
6 Phase 2 viewed from phase 1
7 View towards amphitheatre with the cathedral beyond

4 场地平面图
5 透过港口码头看到的一期工程外景
6 由一期工程看到的二期工程外景
7 圆形露天广场外景，旁边为教堂

Offices for Lloyds Bank PLC 171

8

9

8　二期建筑南侧
9　二期建筑码头区一侧
10　一期建筑端部的阶梯型山墙
11　一期建筑中的室内走廊
12　主入口门厅

8　Phase 2 looking south
9　Phase 2 quayside
10　Phase 1 end gable stair
11　Phase 1 Galleria
12　Main entrance hall

Offices for Lloyds Bank PLC　173

13

14

15

13 Route to phase 2
14 Galleria meeting space
15 View into phase 2 courtyard
16 Phase 2 courtyard

13 通往二期建筑的道路
14 走廊之间的聚会场所
15 二期建筑庭院外景
16 二期建筑庭院

16

Offices for Lloyds Bank PLC 175

Copthorne Hotel

Design/Completion 1989/1991
Newcastle-upon-Tyne
Copthorne Hotels (Newcastle) Limited
16,858 square metres
Post-tensioned prestressed concrete flat slabs, steel roof frames
Cast stone, steel and glass

科坡斯恩酒店（Copthorne Hotel）
设计/完成　　1989/1991
泰恩河畔的纽卡斯尔
科坡斯恩酒店（纽卡斯尔）有限公司
16,858m²
后张预应力混凝土平板，钢屋盖框架
预制石料，钢材，玻璃

The Copthorne Hotel occupies an important riverside site on the River Tyne close to the centre of Newcastle. Built alongside a group of Grade 2 listed historic warehouses, the new hotel has been designed to respond to its local setting and the wider context of the waterfront. Framed by the existing bridges over the River Tyne, it provides a total of 156 en-suite bedrooms on five floors above the public areas. In addition to bars, lounges, restaurants and meeting rooms, the design also provides a fully equipped health club with swimming pool. A three-level structure for 200 cars has been integrated within the envelope of the new building.

A strip of ground has been reclaimed from the River Tyne to form a new riverside walk. The building sits astride Newcastle's medieval Curtain Wall and the Riverside Tower. The archaeological works on the site were protected and the new structure designed to span across them.

The main entrance of the hotel is defined by a porte-cochère linked directly to an atrium lounge. The public rooms all have views over the Tyne. The bedrooms also have views out over the river and the city.

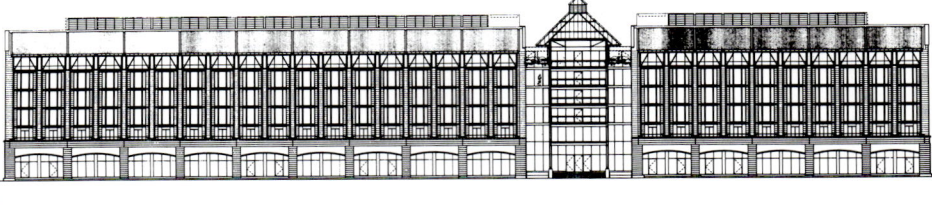
1

2

科坡斯恩酒店坐落在泰恩河畔，位于临近纽卡斯尔市中心的显要位置。酒店附近有许多列为二级保护古建筑的仓库建筑，因此设计认真考虑了与周边建筑环境和河畔景观的协调关系。河上已有的桥梁限定了酒店的建筑风格。酒店首层为公共区域，上部5层共设有156套标准套间。大楼内设有酒吧、休息室、餐馆和会议室，还有一间设施完善的健身俱乐部和游泳池。大楼内还有3层停车场，可供200辆汽车停放。

沿泰恩河畔开辟了一条新的河畔人行道。场地周围的古建筑如纽卡斯尔的中世纪幕墙和河畔塔楼都得到了良好的保护，新酒店就穿越在这些古建筑群之间。

酒店的主入口可供车辆出入，沿此可进入中庭里的休息室。公共区中可以眺望到泰恩河景色，而在楼上的酒店套房里能够俯瞰泰恩河和整个城市的景色。

1　South elevation
2　The building is framed by existing bridges
3　Study model of riverside facade
4　Interior view of the swimming pool in the leisure complex
5　The new hotel fronts the River Tyne and defines a new waterfront

1　南立面
2　已有的桥梁限定了酒店的建筑风格
3　建筑临河立面模型
4　休闲会所中的游泳池内景
5　新酒店位于泰恩河畔，营造出新的河畔景观

3

4

5

Copthorne Hotel 177

123 Buckingham Palace Road

Design/Completion 1988/1990
Victoria, London SE1
Greycoat London Estates Limited
51,746 square metres
Reinforced concrete
Stone paving, glass skin, steel, masonry wall

This project presented Arup Associates with the opportunity to plan a development over the main line railway into Victoria Station in the centre of London. The resulting low-rise building on Buckingham Palace Road incorporates a new bus station and shops at street level with six floors of offices above. An existing historic screen wall along the street was retained and shops, cafes and restaurants were sited along a covered pedestrian street and around a new square to link the site into the fabric of the city. The steel-framed structure, detailed to allow the building to be built without disrupting rail services, was commended in the 1992 Structural Steel Awards.

Continued

在这一项目中，阿鲁普联合事务所在伦敦市中心的维多利亚火车站周边地区沿着铁路干线进行了规划开发。工程在白金汉宫大道临街建造了一批低层建筑，首层为一家公共汽车站和一批商店，上部6层则是写字楼。街道中的古幕墙得到了修复，商店、咖啡厅和餐厅围绕着人行街道和新建的广场布置，人行街道建造了天篷，广场的设计也符合城市布局的需要。建筑采用了钢结构，从而使施工过程不会影响该地区的轨道交通，为此工程还获得了1992年的钢结构奖项。

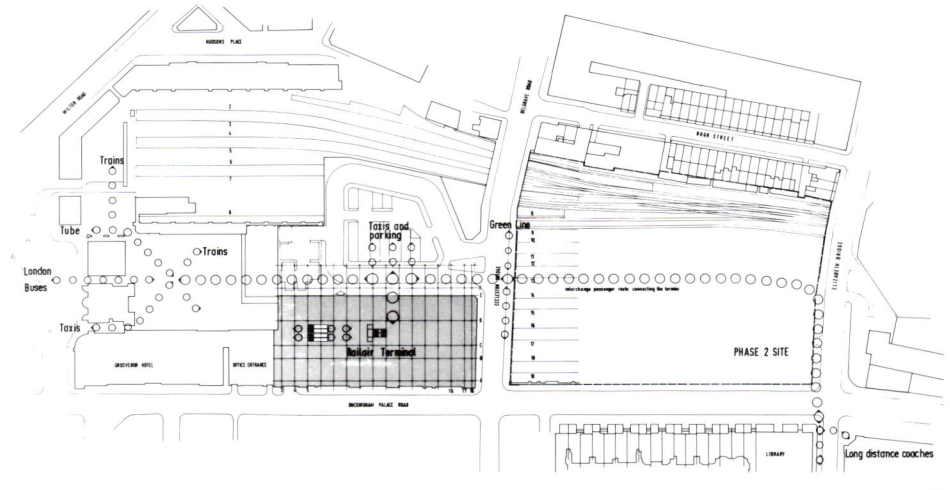

1

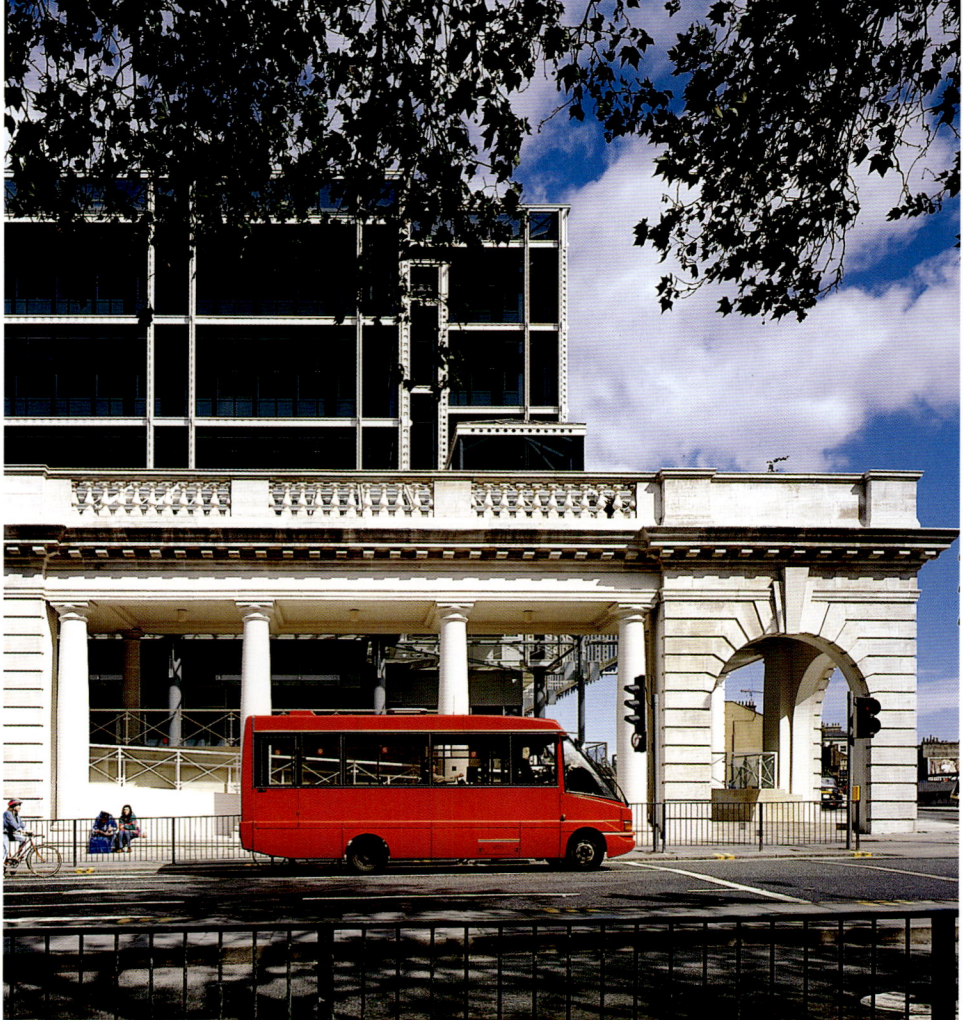

2

1 Site plan
2 The historic stone wall to the street has been retained to create a screen to the new building
3 A glazed court at the heart of the scheme provides a new public space in the city

1 场地平面图
2 沿街而建的古石墙得到了修复，并作为新建筑的屏障
3 建筑的中心设有玻璃顶棚广场，可作为公共活动空间

The offices for the PA Consulting Group within this new building accommodate the reception, lifts and access bridges to each of the office floors within a central day-lit atrium. A restaurant for the staff overlooks the atrium. On each floor there are both open and enclosed offices together with reception areas, conference rooms and presentation suites. The design proposals included office space planning, interior design and the design of special furniture and fittings including customised ceiling tiles which incorporate the light fittings.

PA咨询集团公司（PA Consulting Group）的办公室就设在新的写字楼里。新楼中设计了中央天井，中央天井采用顶部自然采光。天井中设有接待区、电梯和天桥，从此可去往每个办公室楼层。在员工餐厅里，可以俯瞰天井中的景色。在每个楼层，办公室按照需要都分为开放式办公区、独立办公室、接待区、会议室和演示室等。设计方案还包括办公楼内部空间规划、室内设计和特殊的设备家具的设计，包括用户自定义的顶棚瓷砖和灯光照明效果等。

4

4　The central day-lit atrium
5　A restaurant opens onto a day-lit atrium in the offices of the PA Consulting Group
6　Interior of the PA Consulting Group offices overlooking atrium galleries which provide access to the lifts

4　中央天井，采用自然采光
5　PA咨询集团公司员工餐厅，由此可以俯瞰天井中的景色
6　PA咨询集团公司办公室，可以俯瞰到天井四周的廊道，廊道通往电梯

Horsham Park, Genshagen

Appointed 1991
Genshagen, Germany
Horsham Properties PLC
16,500 square metres
Steel frame
Glazed curtain wall with solid panels and external sunscreens

霍莎姆园区，盖莎根

中标时间　　1991
盖莎根，德国
霍莎姆投资开发公司
16,500m²
钢框架
玻璃幕墙，实心面板，遮阳帷幕

Arup Associates have designed the first three buildings for a new business park at Genshagen which is located on the southern side of Berlin alongside the Berliner Ring. Situated on prominent sites close to the entrance to the park, the buildings have been designed to provide for a variety of different tenancies and to accommodate a mix of both office and light industrial uses.

The two- and three-storey buildings range in size from 5,000 to 6,500 square metres. They have been designed as pilot projects which test and develop the traditional German Building Regulations. The designs are energy-conserving, using natural ventilation, highly insulated building envelopes and sun-shading to glazed areas. Mechanical ventilation is used only for extremes of temperature.

1

　　这个新建的商业园区位于柏林南部的盖莎根，阿鲁普联合事务所承担该园区的第一批三幢建筑的设计。这三幢建筑位于园区入口的显要位置，设计要求建筑满足不同租户的需要，可同时容纳办公室和轻工业厂房。

　　建筑多为2－3层，面积由5,000m²至6,500m²不等。这些建筑作为先驱者，发展和检验了传统的德国建筑设计规范。建筑设计秉承节能的原则，采用自然通风，配备高隔热的围护系统，玻璃幕墙采用遮阳帷幕。因此，仅仅在特别必要的温度条件下才需要采用机械通风设备。

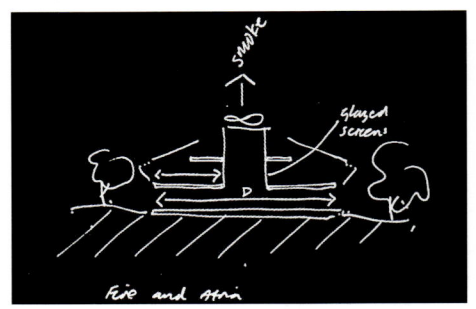
2

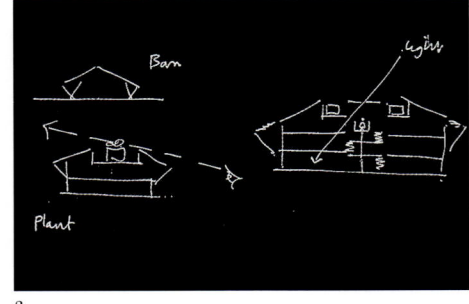
3

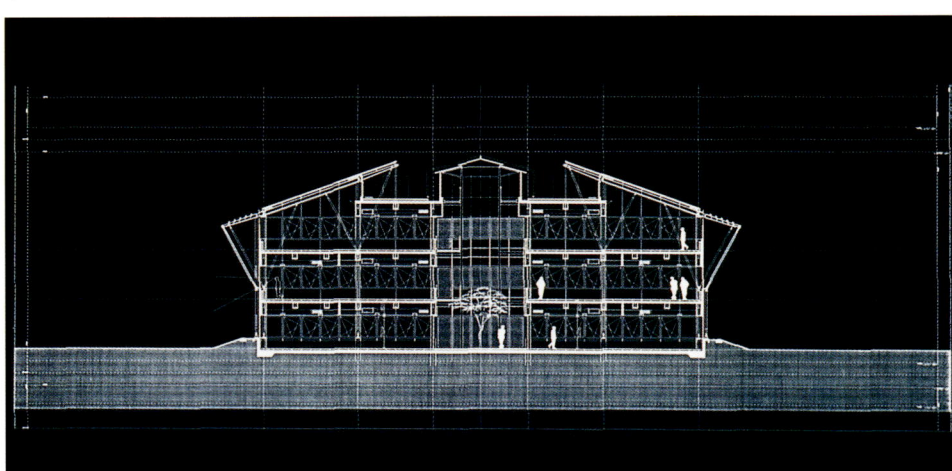
4

1 Site model
2 Initial design study for ventilation
3 Initial design studies for daylighting
4 Scheme design: typical section for artificial lighting
5 Typical elevation
6 Study model

1 场地模型
2 通风设计初步方案
3 采光设计初步方案
4 方案设计：人工照明系统典型剖面
5 标准立面
6 建筑模型

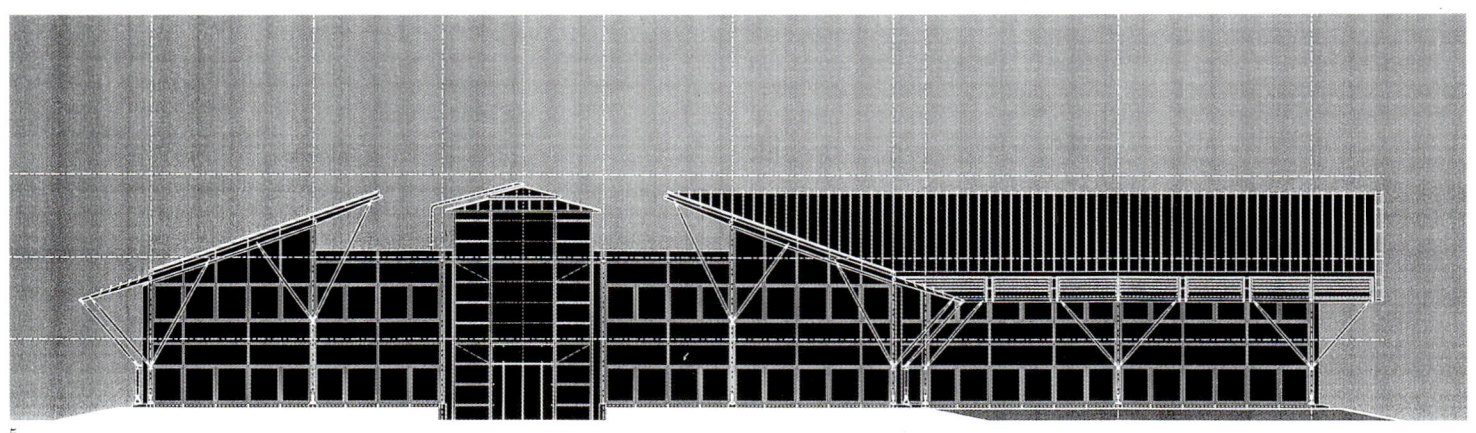

5

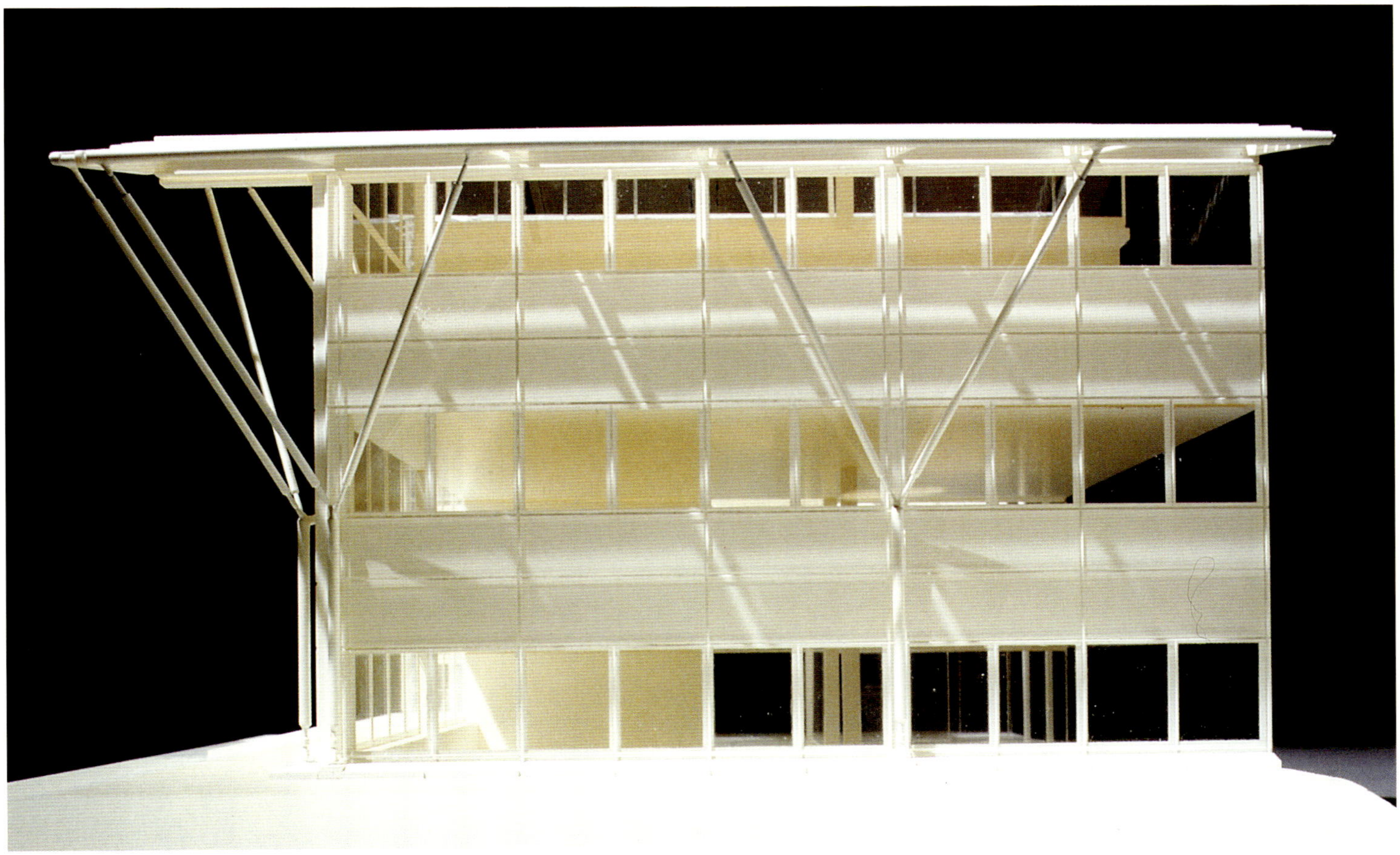

6

Horsham Park, Genshagen 183

Jubilee Line Extension Service Control Centre

Appointed 1991
Neasden, London NW10
London Underground Limited/Jubilee Line Extension
3,500 square metres
Concrete cladding panels, metal cladding and glazing

朱比利地铁线附属服务控制中心

中标时间　　1991
尼斯登，伦敦西北10区
伦敦地铁有限公司/朱比利地铁线附属设施
3,500m²
混凝土盖板，金属包层，玻璃

The new Service Control Centre will function as a combined facility for the Jubilee, Metropolitan, Circle and Hammersmith and City lines in London.

The building's form is articulated into two distinct elements: a served element, including the control room, equipment room and administrative area, which is semi-circular in shape; and a serving element consisting of support spaces and plant housed in a rectangular block.

The building is three storeys high and will be situated against an existing earth embankment. This will be excavated to provide access at road level while reducing the mass of the building from the residential gardens nearby.

The whole composition, with retaining walls and security lodge, encloses a secure inner compound for service vehicles and parking. A louvred facade onto the compound gives access and air flow to the building's plant areas.

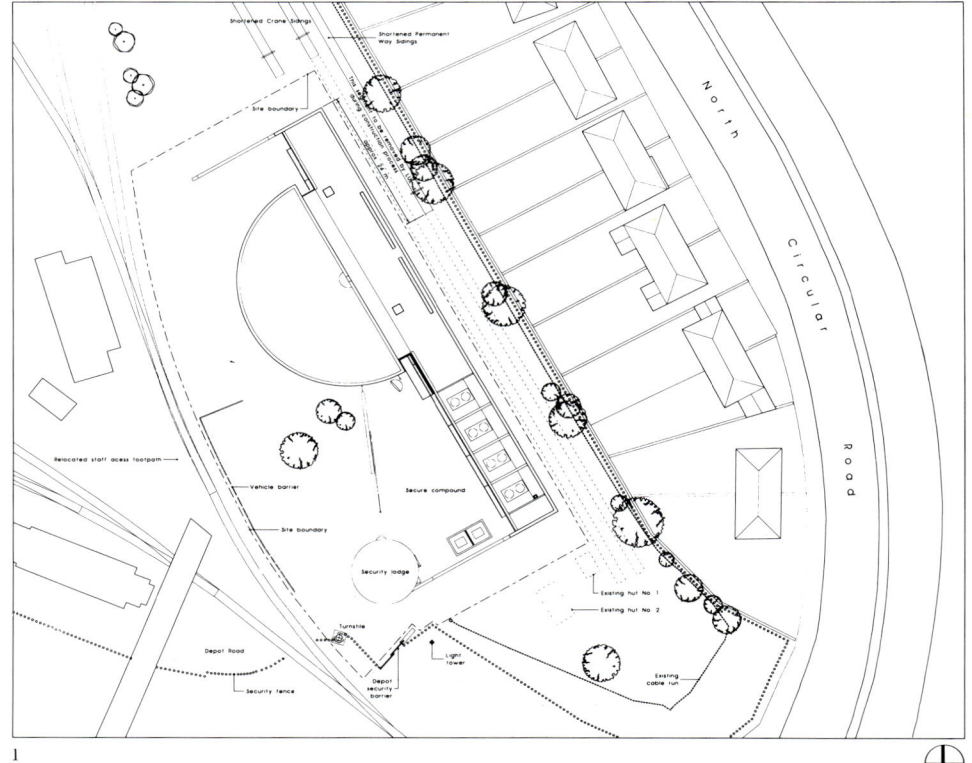

1

Scale 1:500

2

该项目将成为伦敦几条相交叉的地铁线的中心枢纽，包括朱比利线，大都会线，环线，锻工线和城市线等。

建筑分为两个连通的独立单元，一个单元是半圆形的，其中设置了控制室、设备室、管理区；另一个单元则是长方块，内设供应中心和厂房。

大楼为3层，倚着原有的路基土堤而建，同时开挖了通往平地道路的通道。这样，从附近的住宅区的视角看去，建筑的体量就不再显得过于庞大。

建筑外围还设有围护墙和门岗，围出的场地作为停车场使用。建筑正面和入口就面朝围场，正面墙上设有通风口，使建筑可与场中绿化区域实现通风。

1	Site plan	1	场地平面图
2	Scheme design model	2	方案设计模型
3	Second-floor plan	3	三层平面图
4	Scheme design model	4	方案设计模型

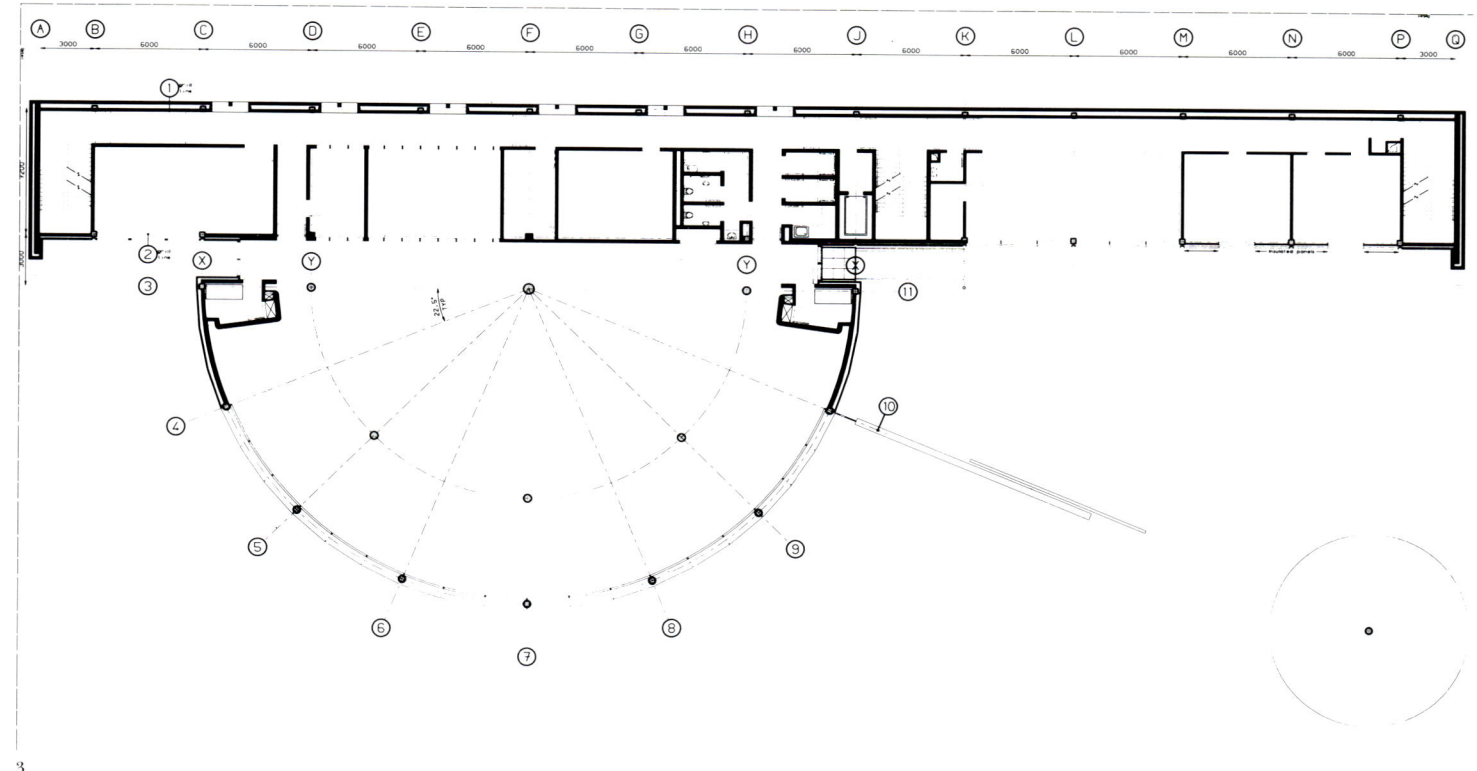

Jubilee Line Extension Service Control Centre 185

Royal Insurance House

Design/Completion 1987/1991
Peterborough, Cambridgeshire
Royal Life Holdings Limited
20,530 square metres
In situ concrete skeletal frame, steel frame
Glazed wall with aluminium and stainless steel framework; brick

皇家保险大楼

设计/完成　1987/1991
彼得伯勒，剑桥郡
皇家人寿保险资产有限公司
20,530m²
现浇混凝土框架，钢框架
玻璃幕墙，采用铝、不锈钢框架；砖

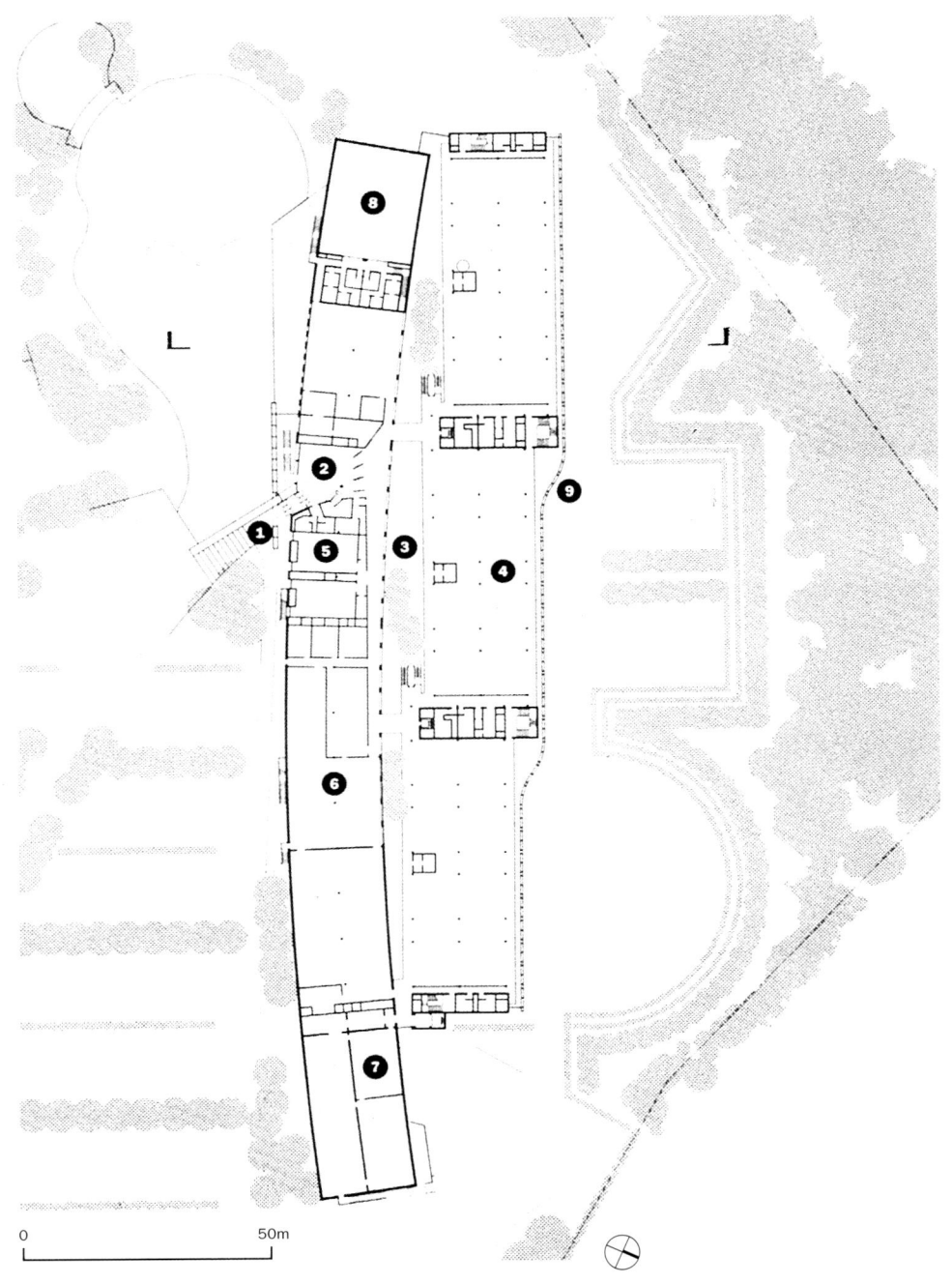

The new offices for Royal Life Holdings Limited provide accommodation for up to 1,000 staff together with a large computer suite, training rooms, restaurant facilities and a sports club. There is parking on the site for 750 cars.

The 7.2-hectare site at Lynch Wood is on the outskirts of the city, and the design emphasises this boundary between town and country.

The low-rise building provides office work spaces in three linked blocks of three floors, and has direct views to a series of newly created gardens and the open countryside beyond. Planned as a group of "outdoor rooms", the gardens are defined by enclosing beech hedges, landscaped banks and lime trees.
The glazed north wall of the offices has a free-standing white trellis framework which becomes an element within the gardens and which relates to both the internal and external spaces. It also accommodates opening lights for ventilation to the offices.

Continued

此项目为皇家人寿保险资产有限公司的新办公楼，可以容纳1000名员工办公，同时也设有大型计算机房、培训室、餐厅设施和一间运动俱乐部。大楼还配备了可停放750辆轿车的停车场。

场地位于市郊的林奇·伍德，占地7.2hm²，设计着重考虑了建筑所处的市镇与乡村交界的地理位置的特点。

大楼为3层建筑，由三个连通的楼区组成，从楼内的办公空间可以直接眺望新建的花园和远处开阔的乡村景色。花园采用山毛榉树篱、菩提树和特别设计的土堤围护，试图营造一组"户外房间"。办公楼北面的玻璃幕墙采用独立的白色方格框架，恰到好处的将内部和外部空间相连通，从而成为了花园的一部分。玻璃幕墙中设有照明设备和通风系统。

1

1	Main entrance	6	Computer suite	1	主入口	6	计算机房
2	Reception	7	Central plantrooms	2	接待区	7	中心设备房
3	Internal street	8	Sports hall	3	内部街道	8	运动馆
4	Office areas	9	Glazed screen wall	4	办公区	9	玻璃幕墙
5	Training rooms			5	培训室		

1	Main entrance
2	Reception
3	Internal street
4	Office areas
5	Support facilities
6	Service tunnel
7	Glazed screen wall
8	Steel roof
9	Formal gardens

1	主入口
2	接待区
3	内部街道
4	办公区
5	设备房
6	供应管道
7	玻璃幕墙
8	钢屋盖
9	整齐的花园

1	Site plan
2	Section
3	The south-facing dining room extends out onto the lakeside terrace
4	The curved enclosing south wall reads as a landscape element
5	The curved wall of the support facilities and the edge of the office decks define the volume of the internal street

1	场地平面图
2	剖面图
3	餐厅面朝南方，延伸至湖畔的露台
4	南面蜿蜒的墙体构成景观的元素
5	办公区与配套设施的外墙间隔出内部的街道

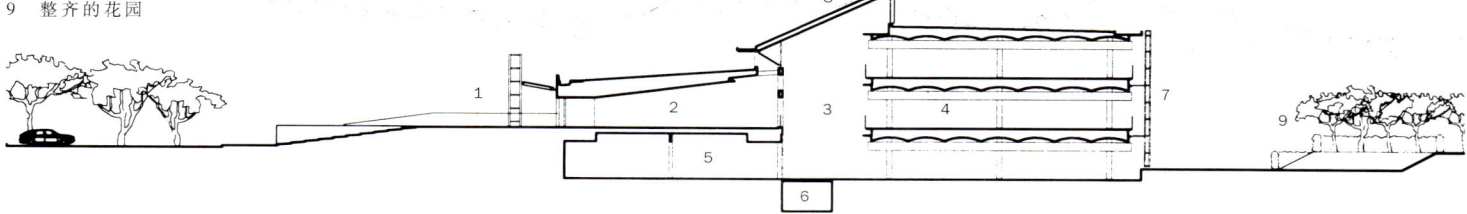

2

3

4

5

Royal Insurance House 187

6   The main entrance is marked by the white glazed screen which makes reference to the glazed screen of the north elevation
7   At night the glazed north wall glows with the illumination from the office areas
8   The glazed screen is held away from the slab edges to emphasise its character as a free-standing element
9   The twisting geometry of the roof resolves the radial geometry of the south block with the orthagonal geometry of the office floors

6   白色玻璃幕墙为主入口的标志，参考了北立面的玻璃幕墙的设计
7   北面的玻璃幕墙在办公区的灯光下闪耀着光芒
8   玻璃幕墙和楼板是分离的，以突出它作为独立单元的特性
9   屋顶扭曲的几何体型解决了建筑南区辐向外形和办公区垂直体型之间的协调问题

6

7

Landscaped car parking areas and a main entrance on the southern side of the site are screened from the work spaces and the countryside beyond by a wall which encloses the office support facilities. These two sets of spaces are linked by a top-lit "street".

The offices have been carefully designed to conserve energy. The design provides high levels of natural daylight and the orientation reduces heat gain from solar radiation and minimises glare. The scheme also provides a wide range of facilities for staff including a south-facing restaurant overlooking a new lake, a sports hall, lounges and outdoor sports fields.

场地南部设有主入口和停车场，建筑物及周围的树木对停车场有遮荫作用。南面的墙内是办公楼的配套设施建筑，配套建筑与办公楼之间由一条内部的"街道"相连，街道采用顶部自然采光。

办公楼按照节能的原则进行了精心设计。设计大量采用了自然采光，建筑的朝向考虑了减少日照和射线作用的需要。方案还为员工设计了一系列的配套设施，包括朝南开放的一间餐厅，从餐厅可以眺望新修的人工湖；另外还有一间运动馆、休息室和户外运动场地等。

8

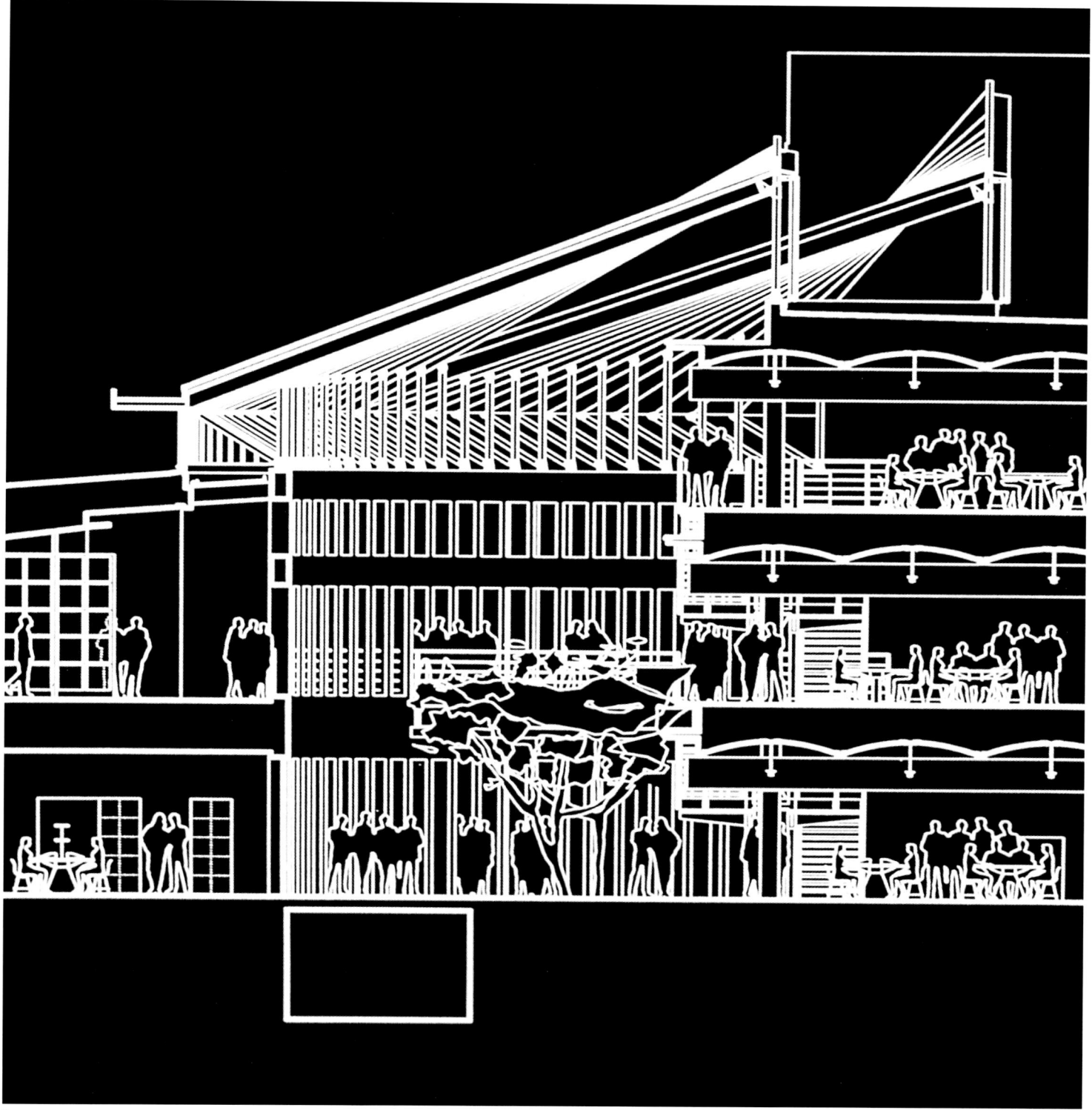

Berlin 2000 Olympic Facilities

Competition 1992
Berlin, Germany
Stanhope Properties PLC
180,000 square metres
Reinforced concrete and steel

柏林 2000 年奥运会场馆

设计竞赛时间　　1992
柏林，德国
斯坦诺普投资开发公司
180,000m²
钢筋混凝土和钢材

The design was prepared as one of three selected projects considered by the Berlin Senat in Germany, for the Berlin submission to host the Olympic Games in the year 2000.

It is based on a master plan for the development of a 13.6-hectare site (currently occupied by a run-down football ground) on Chaussestrasse—an extension of Friedrichstrasse in Berlin. The proposal was to create a new district with residential development, hotels, shopping and office space totalling 180,000 square metres, together with a 15,000-seat Olympic stadium and a 5,000-seat warm-up hall. The major Olympic events which the stadium was designed to accommodate include gymnastics, volleyball, handball, basketball, indoor athletics and ice hockey.

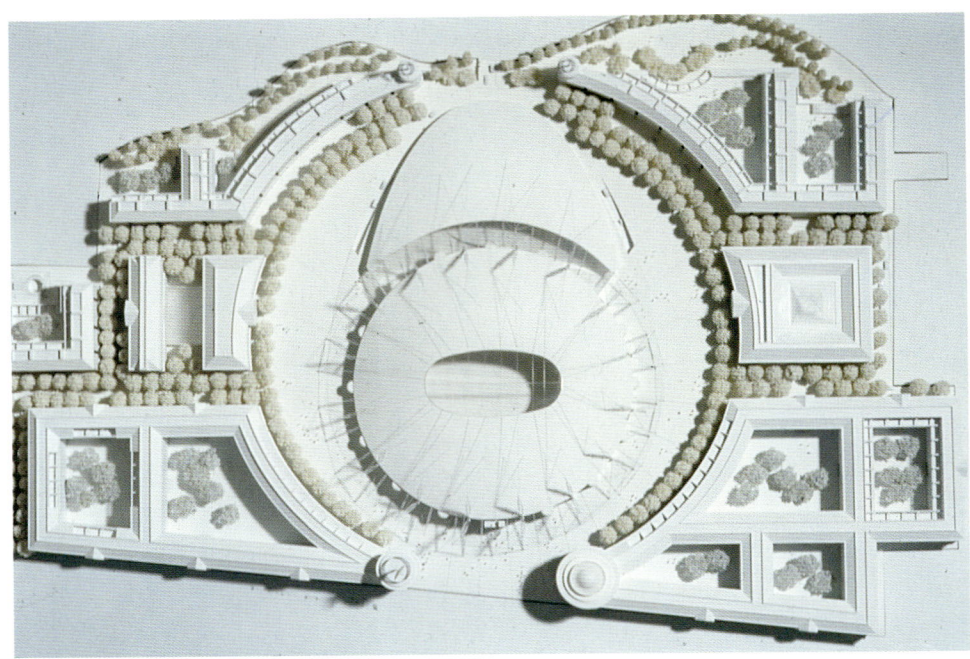

1

该项目为柏林市为申办 2000 年奥运会而规划的三大工程之一。

场地位于柏林的乔叟大街—弗里德里希大街的延伸部分，占地 13.6hm²，当时还是一个破败的足球场。工程计划建造 15,000 座的奥林匹克体育场和 5,000 座的室内体育馆，同时还配套建造面积总共达 180,000m² 的住宅区、旅馆、商店和写字楼。体育场内可举行的奥运体育项目包括体操、排球、手球、篮球、室内运动和冰球等。

2

1　Plan view of block model
2　View over main entrance
3　Plan and section
4　Sketch sections of the stadium
5　View of block model showing street elevation

1　场地建筑模型平面视图
2　主入口视图
3　平面图和剖面图
4　体育馆剖面草图
5　建筑模型临街立面图

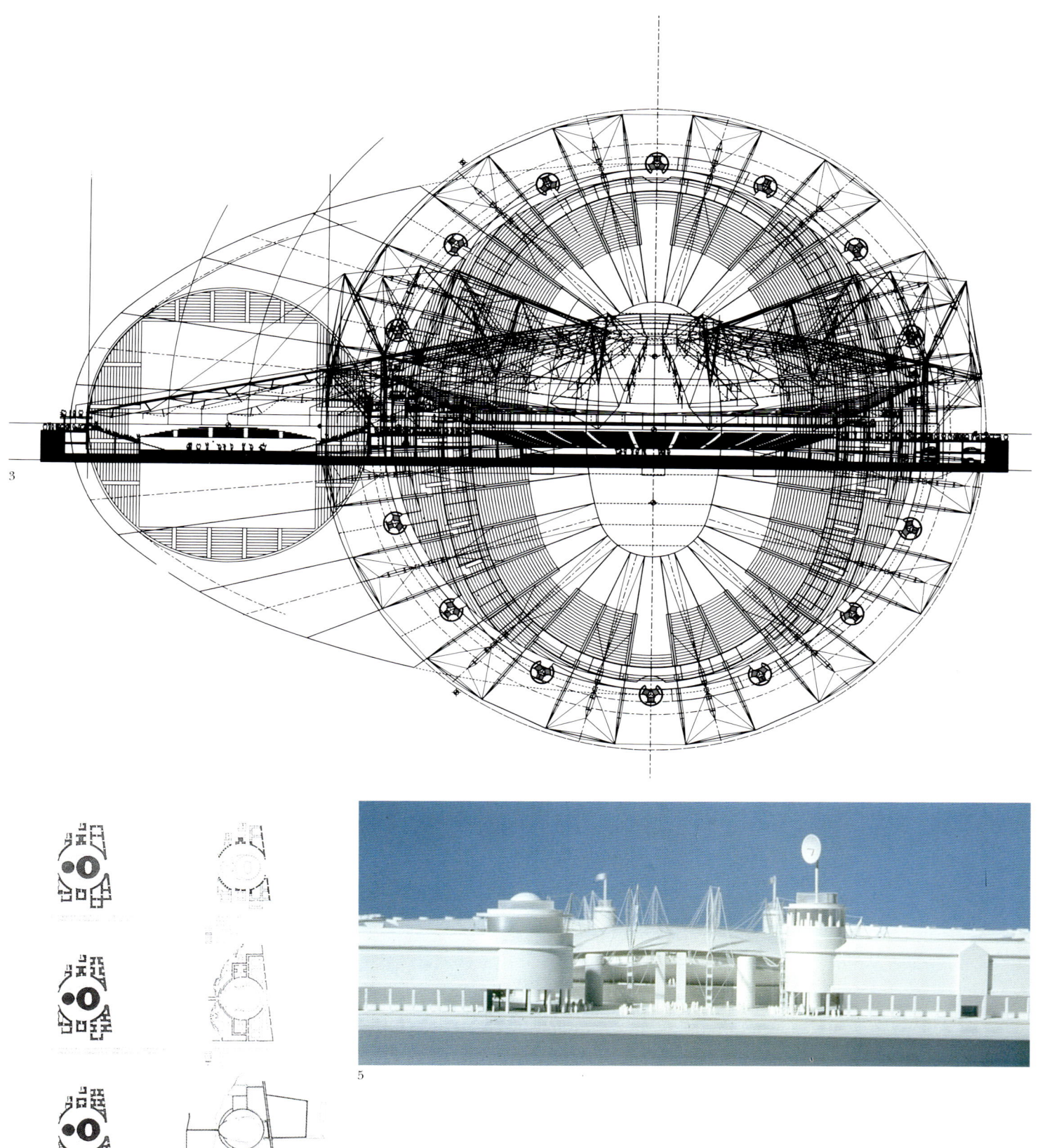

Gonville and Caius College, Cambridge

Appointed 1992
Cambridge, Cambridgeshire
Gonville and Caius College
5,400 square metres (phase 1)
Precast concrete frame
Stone cladding, metal roofs

古维尔和卡尔斯学院，剑桥

中标时间　　1992
剑桥，剑桥郡
古维尔—卡尔斯学院
5,400m² （一期工程）
预制混凝土框架
石块包层，金属屋盖

Gonville and Caius College occupies a site in the West Cambridge Conservation Area. The site includes four fine Victorian villas with mature gardens, and Harvey Court which was designed by Sir Leslie Martin.

A master plan was prepared by Arup Associates for the development of new student rooms and additional social facilities.

The aim of the master plan was to develop a series of residential buildings which defined a route through the gardens, creating new relationships between buildings and landscape while maintaining the intimacy and quality of the existing gardens.

The priorities for the initial phases of development were identified as new undergraduate rooms, a new junior common room, a new multipurpose
Continued

古维尔和卡尔斯学院项目位于剑桥西部保护区内。场地内建有四座维多利亚时代的精巧别墅及花园，以及莱斯利·马丁爵士设计建造的哈维庭院。

阿鲁普联合事务所的任务是在此处规划设计一批新的学生公寓和附属的公共设施。

总规划中住宅建筑的平面布置体现了一种新的建筑物与周围景观之间的关系，楼间的通道就从原来的花园穿过，花园的私密性和品质都得到了保护。

一期工程首先建造的是新的本科生公寓、公共活动室和一间用于举行讲座或戏剧表演的多功能讲堂，另外还包括餐厅等配套设施。古老的哈维庭院也进行了修复和改建。

1

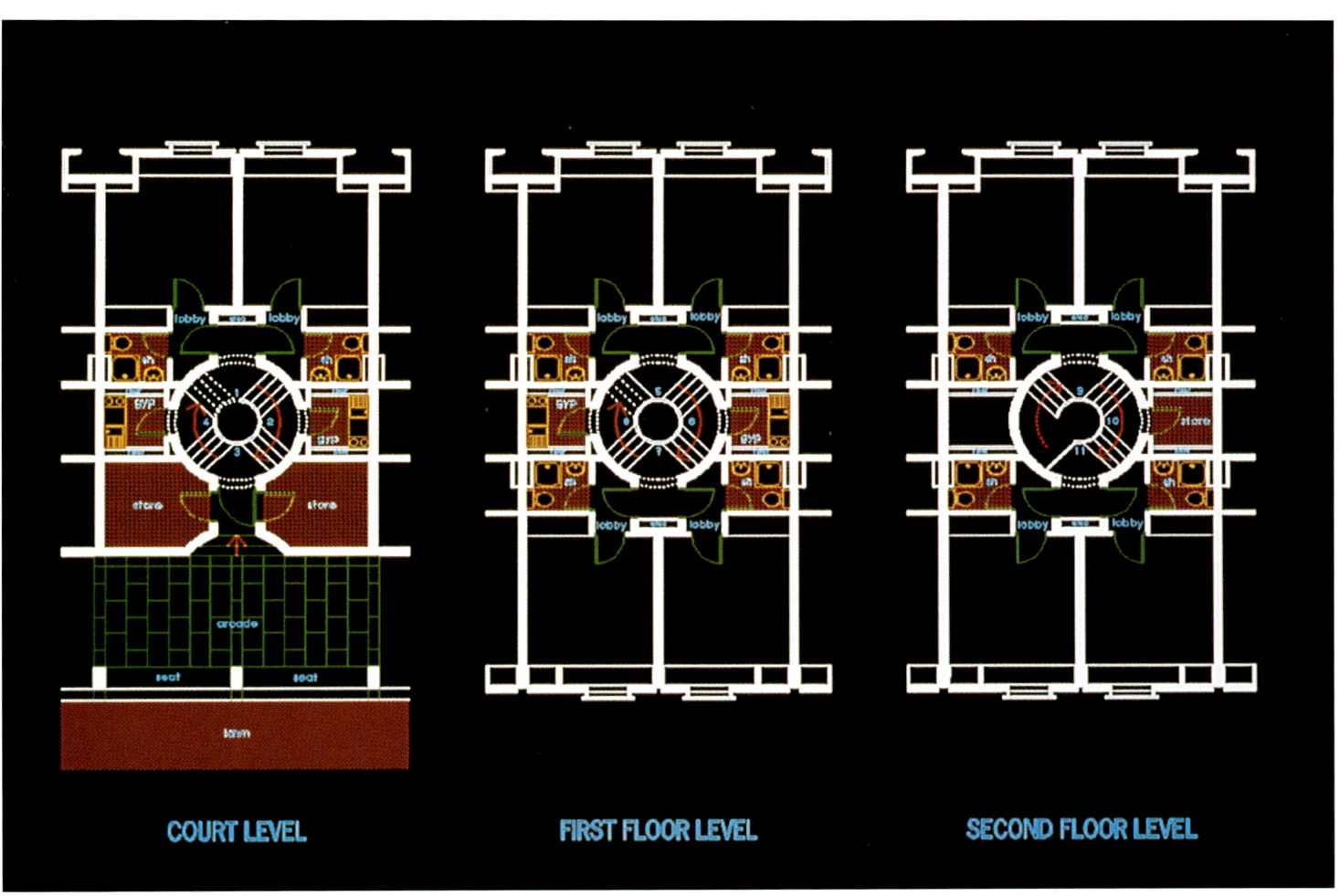

2

1 The master plan site layout allows for long-term development of up to 200 additional rooms
2 Floor plans showing the split level arrangement of rooms off the central staircase

1 总规划平面图，工程在远期将建成另外 200 间公寓
2 楼层平面图，公寓采用夹层结构，楼层和中央庭院中的楼梯对齐

Gonville and Caius College, Cambridge

auditorium designed for lectures and drama, together with alterations to Harvey Court to provide additional dining facilities.

The design for the first phase provides 95 undergraduate rooms adjacent to Harvey Court. The rooms are arranged in a traditional manner, with stairs accessed from an arcade around a grassed court. Beneath this new court is a car park which will replace the existing haphazard parking on the site.

In parallel with this plan, Arup Associates assisted the college in a space study of their Old Courts in central Cambridge.

一期工程毗邻哈维庭院建造了95间本科生公寓。公寓房间的布局遵循传统的样式，楼内设有绿化的中央庭院，庭院四周布置拱廊和楼梯入口。新的庭院的地下设有停车场，用以取代原来的随意停车。

进行这项设计的同时，阿鲁普联合事务所还协助学院对他们在剑桥中部的旧建筑进行了考查和调研。

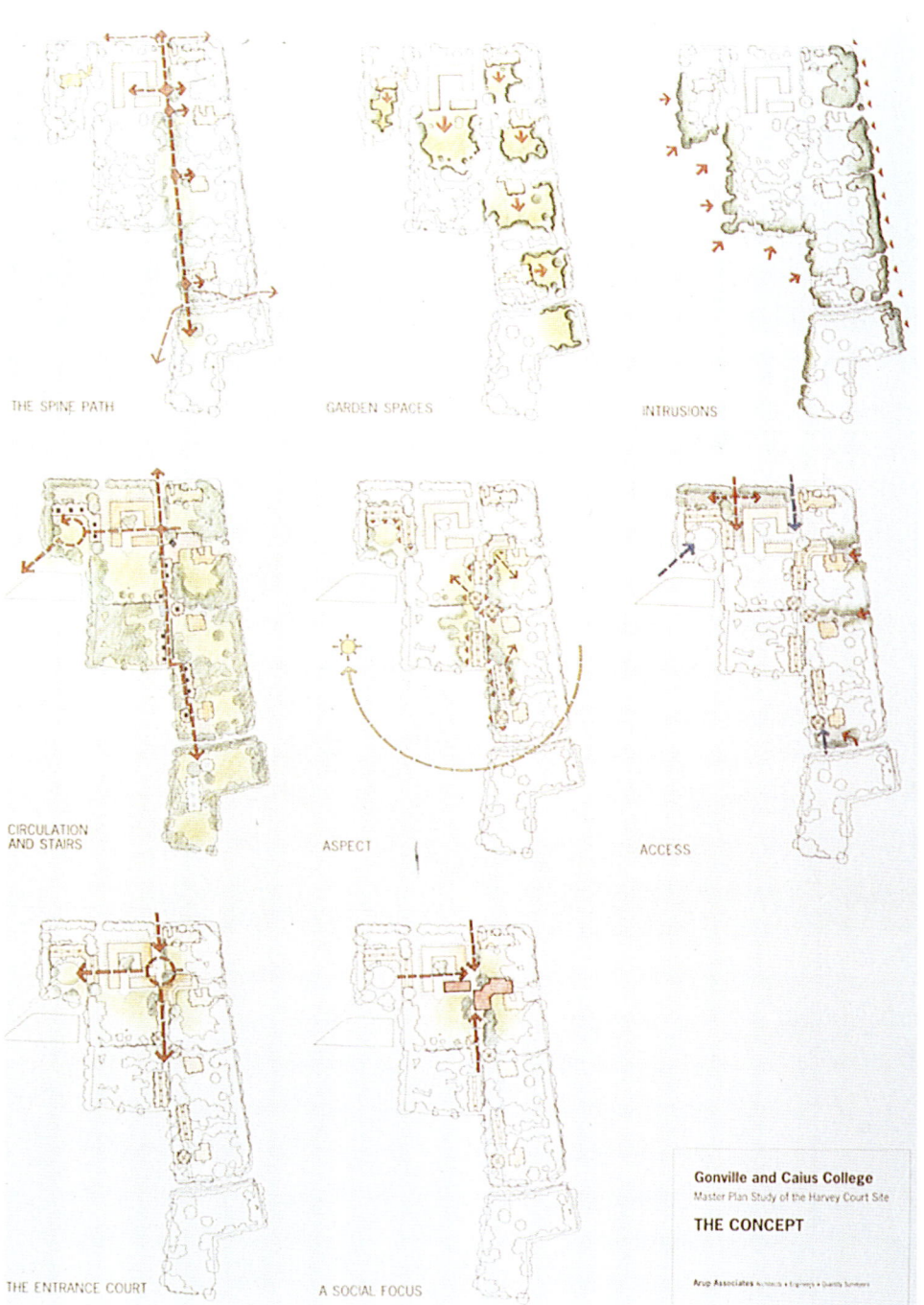

3 Aspects of the master plan concept
4 View of the model showing the south range of seminar rooms and the private residential court beyond
5 Photomontage showing the building as a suburban block within the existing landscaped street frontage

3 总规划原理示意图
4 模型视图，可看到研究室的南侧和旁边的私人住宅院落
5 照片剪辑，建筑兼有城市和乡村特点，毗邻的街道进行了美化

4

5

Gonville and Caius College, Cambridge   195

Hong Kong Central Station, Lantau and Airport Railway

Appointed 1992
Hong Kong
Mass Transit Railway Corporation
415,896 square metres
Reinforced concrete and steel frame
Glazed external walls

香港大屿山－机场铁路中央车站

中标时间　　1992
香港
公共交通铁道公司
415,896m²
钢筋混凝土，钢框架结构
玻璃外墙

The new Hong Kong Central Station will be the terminus of the proposed Airport Railway which will link the city to the new airport and the new town at Tung Chung.

The station will provide facilities for the Airport Express Line (AEL) and the Lantau Line and will be built on land to be reclaimed from the harbour.

The station consists of arrival and departure halls with platforms below ground level. In-town check-in facilities for the airport are planned in the AEL entrance hall at street level. This will allow airport passengers to check in before boarding the express train to the airport. The AEL entrance hall has been designed as a light, airy space with mezzanine levels for concessions and amenities. It also creates a new gateway to the city.

The design also includes a master plan for a 420,000 square metre development above the station consisting of extensive retail accommodation, three tall office towers and two luxury hotels with carefully landscaped public open spaces.

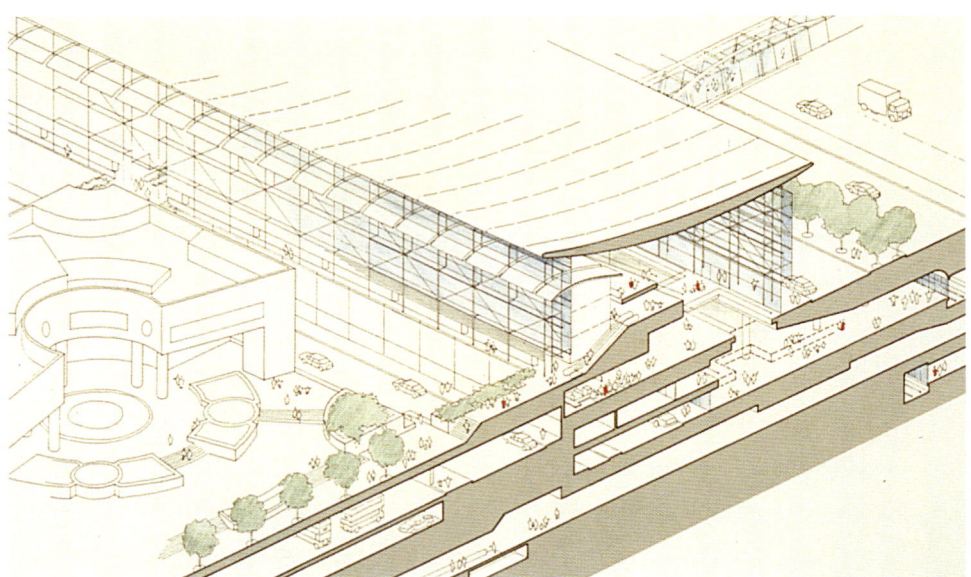
1

新的香港中央车站是规划中的机场铁路的终点，这条铁路将连接市区与新机场以及大屿山岛。

车站将为机场快速铁路（AEL）和朗套铁路线提供配套设施，车站场地来源于港口填海围筑的土地。

车站由抵达区和候车区组成，另外还有地平面位置的站台区。在机场快速线的入口，还设置了机场市内签证处，搭乘飞机的旅客可在乘坐去机场的火车之前就办理好登机手续。机场快速线入口大厅宽敞明亮，环境优雅，底楼与二楼之间的夹层设有休闲娱乐设施。车站也因此成为城市的一道新的大门。

设计还包括总共达 420,000m² 的车站上部建筑区，包括附属的商业区、三座高层办公塔楼、两座豪华酒店以及精心规划的公共活动空间。

2

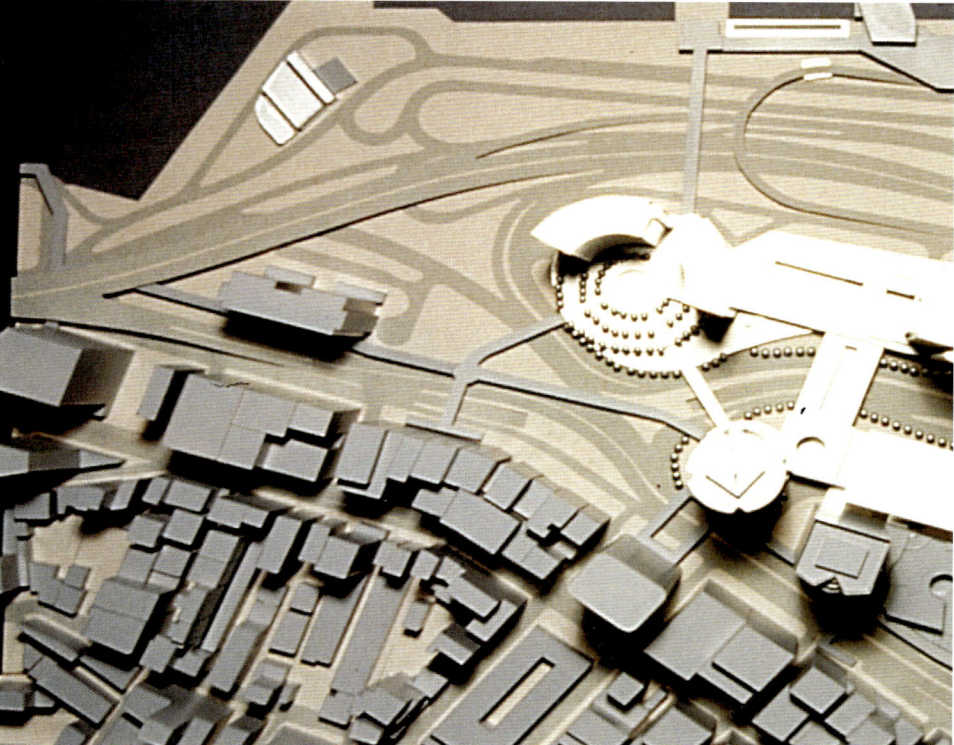

1 Cut-away isometric of the AEL entrance hall and station below
2 Entrance to the AEL departure concourse
3 View of the AEL entrance hall
4 Departure and check-in hall model external glazing study
5 General view of the development from Victoria Harbour
6 Aerial view of model of the development and station on the new reclamation

1 等角投影透视图，图为机场快速线入口大厅和下方的车站区
2 机场快速线候车大厅入口
3 机场快速线入口大厅外景
4 候车区和签证处大厅玻璃外墙模型
5 维多利亚港区域规划模型全景
6 填海地区开发规划模型鸟瞰图

Hong Kong Central Station, Lantau and Airport Railway 197

Nursery School, Sossenheim, Frankfurt

Competition 1992
Sossenheim, Frankfurt, Germany
Stadt Frankfurt am Main
974 square metres
Concrete, timber and blockwork
Glazed with a Kapilux roof

幼儿园，罗森海姆，法兰克福

设计竞赛时间　　1992
罗森海姆，法兰克福，德国
美因河畔的法兰克福
974m²
混凝土，木材，预制砌块

In 1992 Arup Associates were selected as one of six European practices to participate in an international design competition sponsored by the City of Frankfurt to design a school and daycare centre. The design focused on "low entropy", minimising the energy that would be required to build, operate, maintain and eventually demolish the building which is situated within a residential area.

The plan proposed two new buildings sited so that both buildings and outdoor play areas optimise the southern aspect.

The daycare centre is conceived as a brightly illuminated, climate-controlled glass envelope containing a series of classrooms and enclosed by a curving masonry wall.

1

1992年，阿鲁普联合事务所在法兰克福城市举办的有六家欧洲设计公司参与的国际设计竞赛中获胜，获许设计一座幼儿园和日托中心。设计的原则为"节能"，要求建筑在建造、使用、维护及最终拆除的整个生命过程实现能耗最小化。建筑场地位于一片住宅区内。

规划方案包括两幢新建筑，建筑物和户外游乐区域的设计都考虑了南侧景观的优化。

日托中心采用了玻璃围护系统，内设一系列教室。玻璃外壳可使室内获得明亮的光线和良好的温度控制。在建筑外围采用弯曲的砌块墙进行封闭。

1　Three-dimensional computer graphic of model from the east
2　The classrooms are contained within a light glazed envelope and the school is marked by a tower
3　Site plan
4　Model elevation
5　Detail drawing showing the classroom enclosure

1　三维计算机模型，东侧视图
2　教室采用轻质的玻璃围护系统，建有一座作为学校标志的塔
3　场地平面图
4　模型立面图
5　教室围护外墙实景

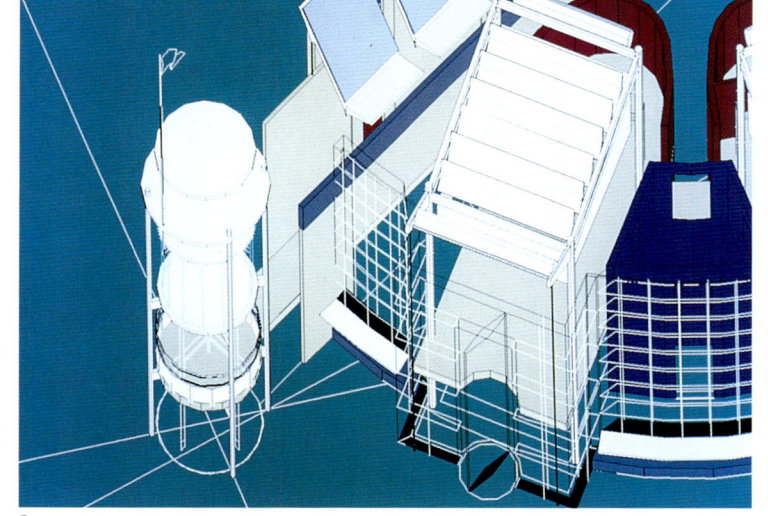

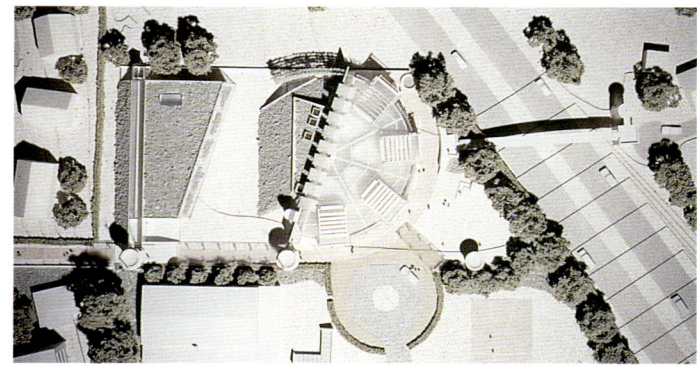

Nursery School, Sossenheim, Frankfurt 199

Düsseldorf Tower

Competition 1992
Düsseldorf, Germany
17,000 square metres
Reinforced concrete, steel and glass

The site for this competition was in the heart of Düsseldorf, immediately above a road tunnel. The design proposed a 20-storey block planned around a raked atrium. The plan of the building has two 12 metre wide wings of offices aligned on either side of the road with the atrium between.

The two wings of offices are linked by cores housing services, main entrance, lifts and stairs. The office floors are raked back with the cores acting as buttresses.

Exhaust fumes and noise from the tunnel precluded openable windows on one face of the building. As this was also the south-facing elevation, the design created a solid facade incorporating solar collectors. The solar collectors are fixed to two towers containing phase-change salts: these store and release energy with small changes in temperature.

Continued

杜塞尔多夫塔楼

设计竞赛时间　1992
杜塞尔多夫，德国
17,000m²
钢筋混凝土，钢材，玻璃

该项目位于杜塞尔多夫市中心，紧接着位于一座公路隧道的上方。大楼共20层，平面布置以一个倾斜的天井为中心，两侧为办公楼，各12m宽。

两翼的办公楼由一系列芯筒相连接，筒内设有服务配套设施、主入口、电梯和楼梯。办公楼内敛倾斜，斜靠着长方形的筒楼。

隧道中的车辆废气和噪声使得建筑南侧不能采用开启式窗户，因此南立面采用了实心的外墙。为了节能，建筑采用了太阳能收集装置，固定在两座塔楼之上，这些收集器带有能量转换器，可以根据温度变化自动调节贮存和释放能量。

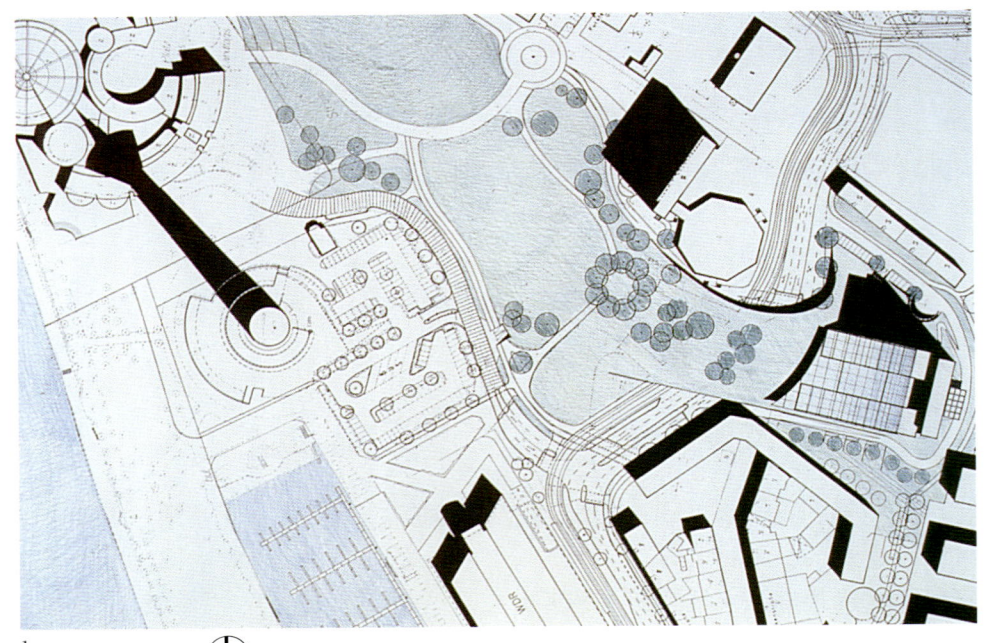

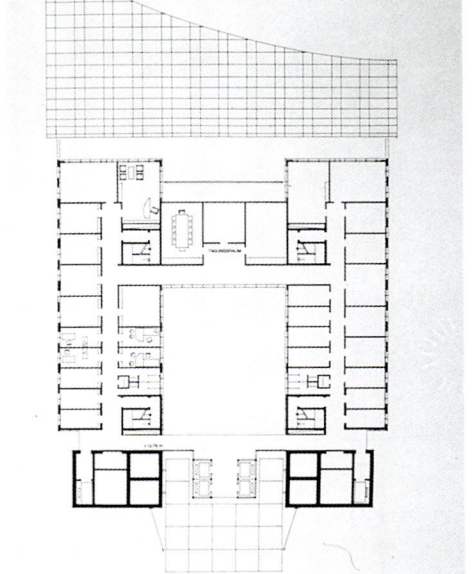

1 Site plan
2 Intermediate plan
3 Long section
4 North elevation

1 场地平面图
2 方案设计图
3 纵剖面图
4 北立面图

3

HEREINFLIESSEN DES PARKS

4

Düsseldorf Tower   201

The building has been designed to be naturally ventilated; opening windows on the east and west are assisted by the stack effect of the atrium. A hollow core construction ensures that cross-ventilation would not be interrupted by a central corridor.

In extreme winter and summer conditions the windows in the offices and cross-flow ventilation openings are closed and a mechanical plenum heating and ventilation system is used. In summer the ventilation plant introduces cool night air through hollow core planks, cooling the building for the following day. In winter the plenum heating plant uses heat from the solar salt store and the co-generation salt store.

建筑采用自然通风，在西面和东面都采用了开启式窗户，天井也能够起到抽吸作用。采用空芯楼板能够保证横向通风不会受内部走廊的阻碍。

在酷热和严寒条件下，办公室的窗户和横向通风将会关闭，而采用机械压力通风采暖系统。在夏季，通风系统可以将夜间凉爽的空气吹过空心楼板，使楼内温度得以降低，以适合白天的需要。在冬季，压力采暖系统将利用太阳能供应暖风。

5

5 Model from north-west
6 Passive solar energy concept
7 Model from west
8 View from offices

5 模型西北侧视图
6 太阳能采集系统示意图
7 模型西侧视图
8 办公楼内视野图

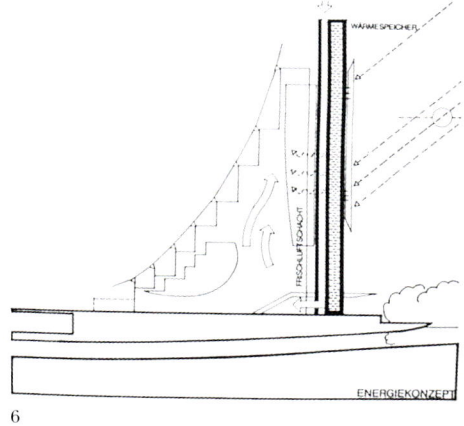

6

7

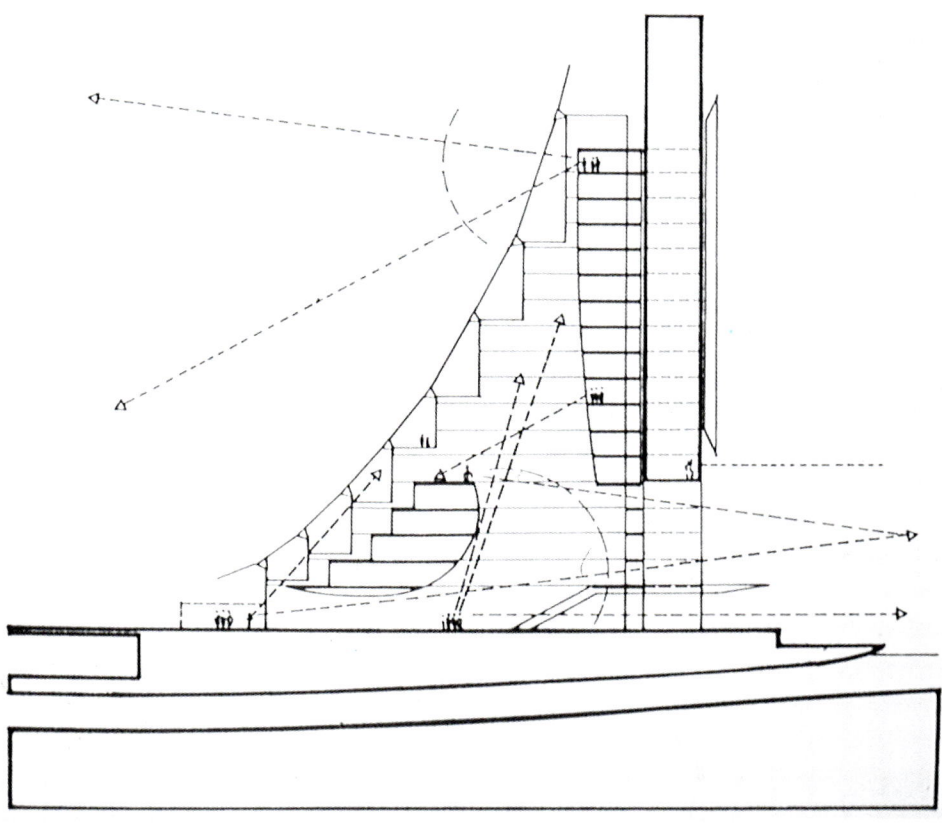

8

Waste-to-Energy Plant

Appointed 1992
Belvedere, Kent
Cory Environmental Limited
23,400 square metres
Reinforced concrete frame, steel framed roof
Profiled metal cladding

废料利用发电厂

中标时间　1992
贝尔威迪，肯特郡
科利环境有限公司
23,400m²
钢筋混凝土框架
钢框架屋盖
异型金属包层

Cory Environmental Limited commissioned Arup Associates to design a waste-fuelled power station on an exposed site beside the Thames, on the south-east side of London. The plant has been designed to relate to the river. The scheme defines a towpath for pedestrians along the riverside and incorporates a jetty to allow deliveries by river. The plant will conserve energy at a neighbourhood scale. Planned on four levels, it has a light, curved skin of silver-anodised aluminium which encloses the process machinery and delivery halls. The form of the building will create a subtle silhouette that will be visible for some distance in the long flat landscapes of the Thames estuary.

The design for the plant was highlighted by the Royal Fine Art Commission as "a superior design of considerable distinction".

该项目中，阿鲁普联合事务所接受科利环境有限公司的委托，设计建造一座采用工业废料作为燃料的发电厂。场地位于伦敦东南泰晤士河的一块无遮蔽区域中。工程沿河边建造了一条人行纤路和一座用于河道运输的码头。发电厂将同比例的贮存能量。建筑分4层，围护结构采用轻质弯曲的镀银面板，生产设备和运输车间设于其中。建筑体现了精巧的外形轮廓，使其与泰晤士河口开阔平直的景观风格相协调。

该项目的设计受到皇家美术委员会的高度赞许，被认为是"十分有个性的精品设计"。

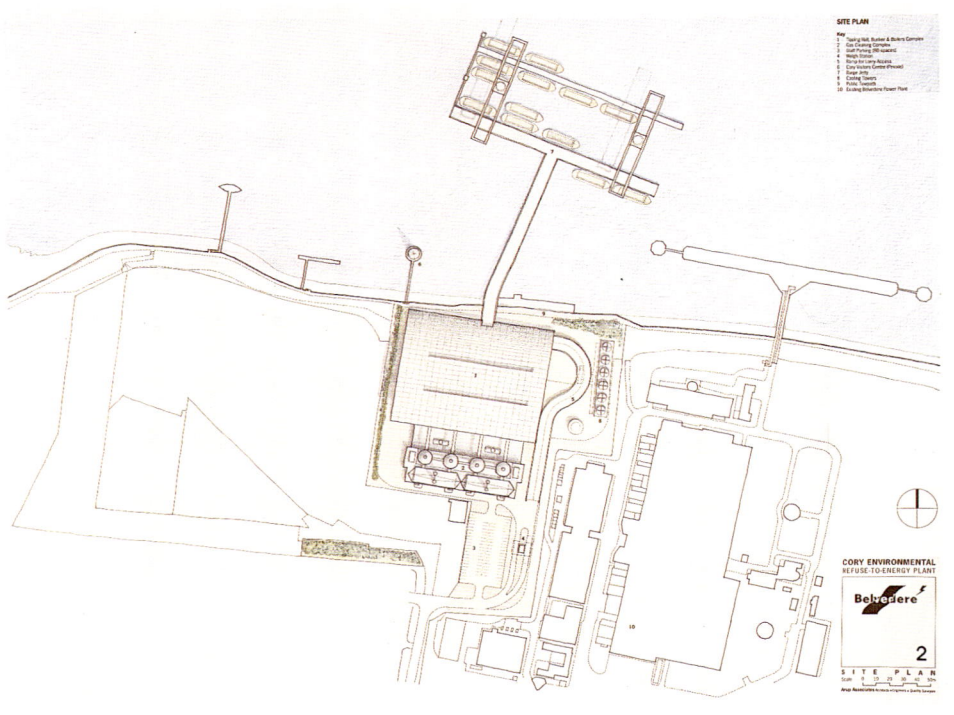

1

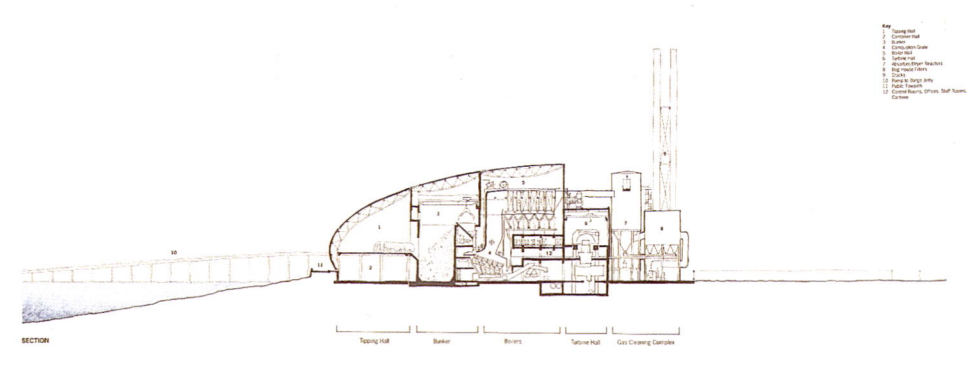

2

1	Site plan	1	场地平面图
2	Section	2	剖面图
3	West elevation and north elevation	3	西立面和北立面
4	View from west	4	建筑西面外景
5	View from east, downriver	5	建筑东面外景，面向河下游
6	View from north-west, from Dagenham across the river	6	由达格南区越过泰晤士河望到的建筑西北方向外景

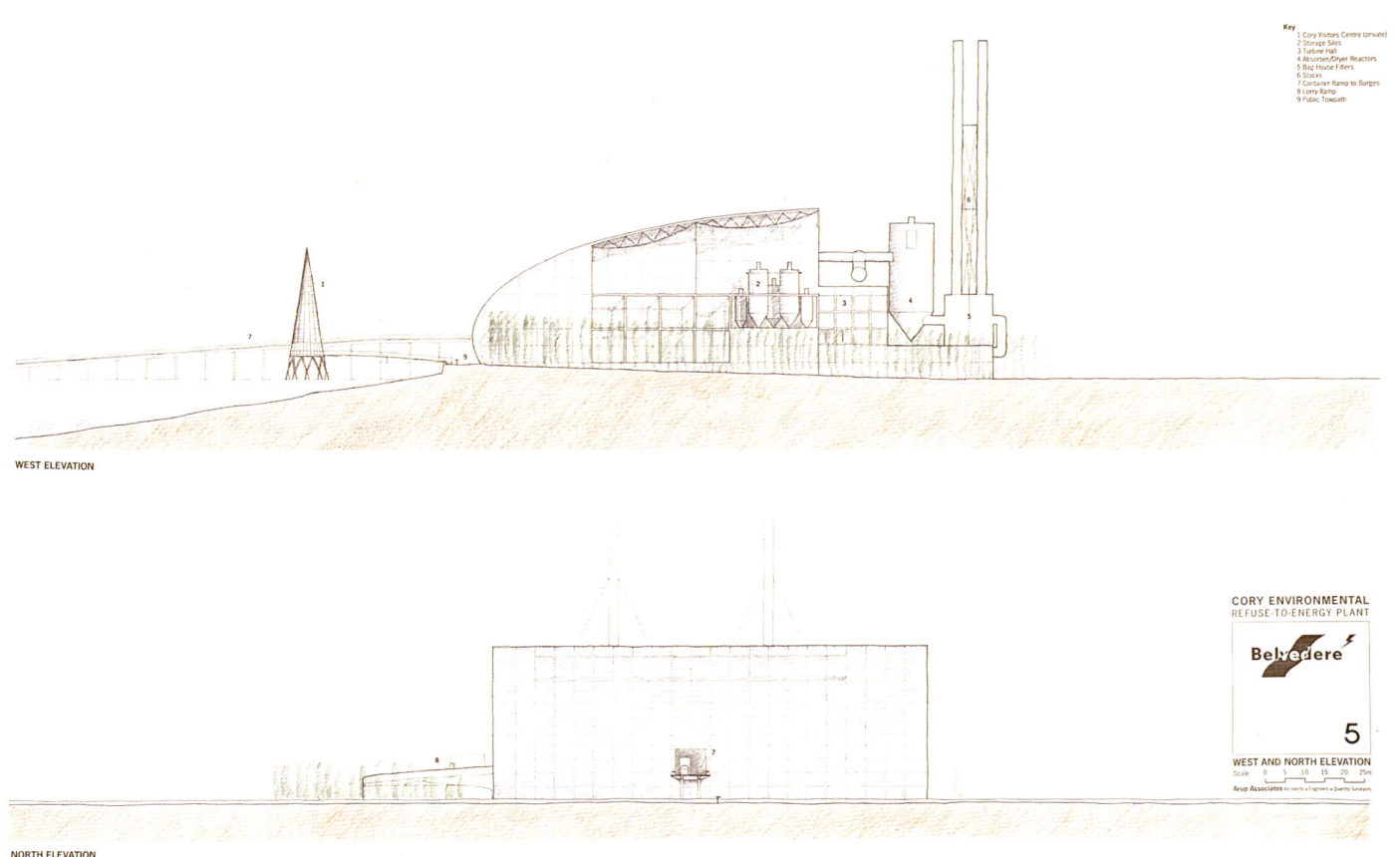

WEST ELEVATION

NORTH ELEVATION

CORY ENVIRONMENTAL
REFUSE-TO-ENERGY PLANT

Belvedere

5

WEST AND NORTH ELEVATION

Arup Associates

3

4

5

6

Waste-to-Energy Plant   205

Grand Lyon Porte des Alpes

Appointed 1993
Lyon, France
Communauté Urbaine de Lyon
250,000 square metres
Site area: 300 hectares

里昂阿尔卑斯科技园区

中标时间　　　1993
里昂，法国
里昂市政署
250,000m²
场地面积：300hm²

The master plan for the new Grand Lyon Porte des Alpes Parc Technologique incorporates buildings for high-technology research and development and for international corporations. These buildings focus on a newly created landscaped park at the heart of the 300-hectare site. This new project will be situated between the city of Lyon and the airport at Satolas.

In addition to the Parc Technologique, the project includes a restructuring of the transportation systems in the area and a major landscaping programme involving new public parks and strategic woodland with open space planting.

阿尔卑斯科技园区作为进入里昂市区的一道大门，建设了一大批建筑，供高科技研发企业和国际性公司使用。园区总占地300hm²，建筑物集中于园区中心地带，建筑区进行了精心的景观美化。这一项目位于里昂市区和萨特拉斯机场之间，成为里昂城的大门。

在建设科技园区的同时，项目还重新规划建设当地的区域交通系统，进行环境改造，包括建造新的公共公园和人造林地。

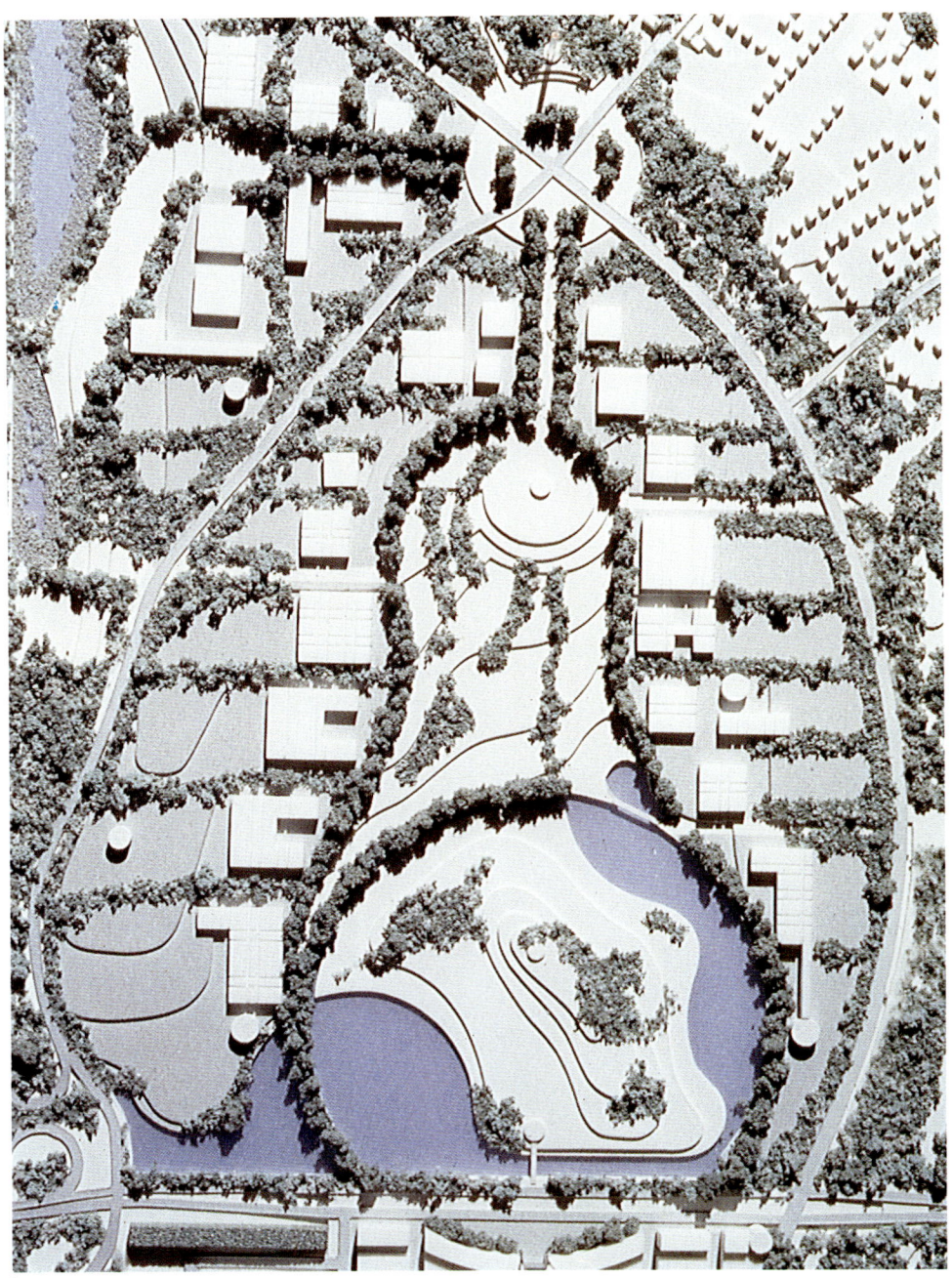

1

1 View of model: Stage 1 Parc Technologique
2 View of model: comprehensive area
3 The site in relation to Lyon and Satolas Airport
4 Landscape principles of Stage 1 of the Parc Technologique

1 模型视图：科技园区一期工程
2 模型视图：全景
3 园区与里昂城及萨特拉斯机场的地理位置关系
4 科技园区一期工程规划示意图

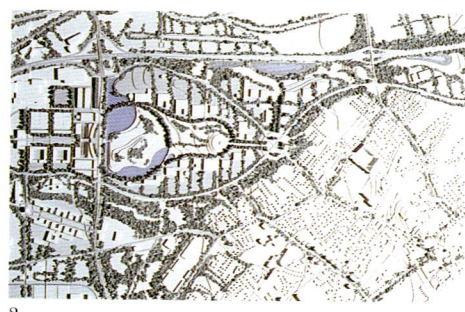

2

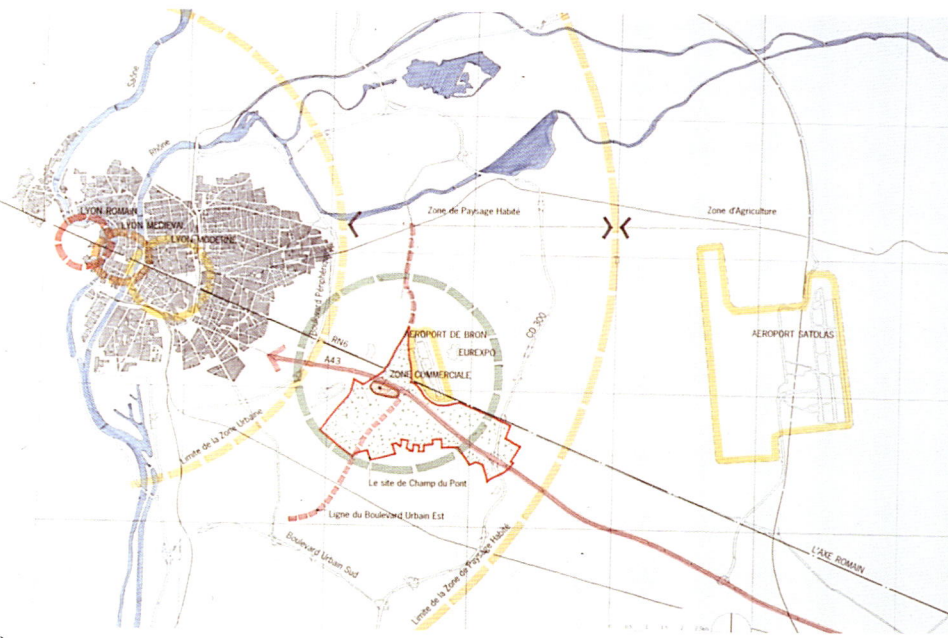

3

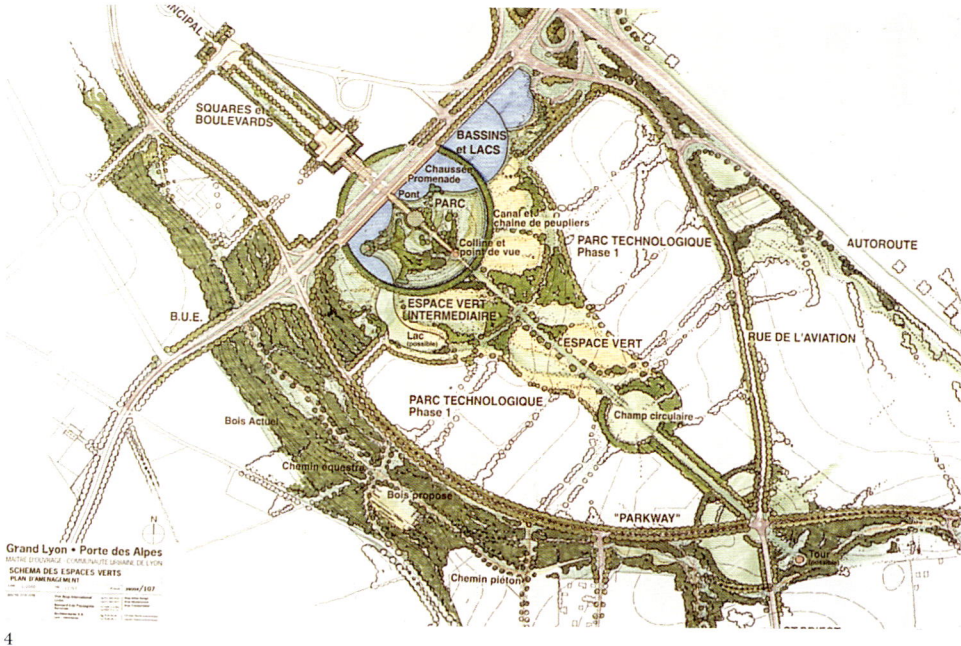

4

Grand Lyon Porte des Alpes   207

Istanbul Cultural and Congress Centre

Appointed 1993
Istanbul, Turkey
Istanbul Foundation for Culture and Art
27,000 square metres
Reinforced concrete frame
Stone cladding, glazing, copper roofs

伊斯坦布尔文化和会议中心

中标时间　　1993
伊斯坦布尔，土耳其
伊斯坦布尔文化艺术基金会
27,000m²
钢筋混凝土框架
石块包层，玻璃，铜屋面

The new Cultural and Congress Centre will be located in a beautiful, sloping wooded area on the outskirts of Istanbul, on land that was formerly part of the Sultan's palace gardens. Two existing historic Ottoman houses together with the original outdoor pool and the tiled pavilion built for the Sultan and his family in the 19th century are still intact, and have been incorporated into the master plan for the development of the new Cultural Centre.

The brief was to provide a symphony hall with 2,500 seats, a chamber music hall for 450, two conference halls to seat 500 and 300 each and a small cinema, with associated restaurant and reception facilities, and parking for 850 cars.

Continued

1

该项目位于伊斯坦布尔市郊的林地之中，场地有一定坡度，周围环境优美，原来是苏丹宫廷花园的一部分。场地中还完整保留着19世纪为苏丹家族建造的一系列古老建筑，包括两座奥斯曼时期住宅、原始的户外水池和瓦顶小亭。项目规划也将这些古建筑考虑在内。

建筑包含一间2,500座的交响乐厅、一间450座的室内音乐厅、500座和300座的会议室各一间和一间小型影院。另外，还设有配套的餐厅和接待设施，以及可停放850辆汽车的停车场。

2

1　Site plan
2　East-west section through main auditorium
3　Revised preliminary design site model, view from north
4　Revised preliminary design site model, view from west
5　West elevation

1　场地平面图
2　东西方向剖面，穿越主会堂
3　修订后场地初步规划模型，北侧视图
4　修订后场地初步规划模型，西侧视图
5　西立面

Istanbul Cultural and Congress Centre 209

6 Second acoustic model
7 Foyer groin vault model
8–9 Model of initial design for concert hall chair

6　二层音乐厅模型
7　门厅穹顶模型
8–9　音乐厅座椅最初设计模型

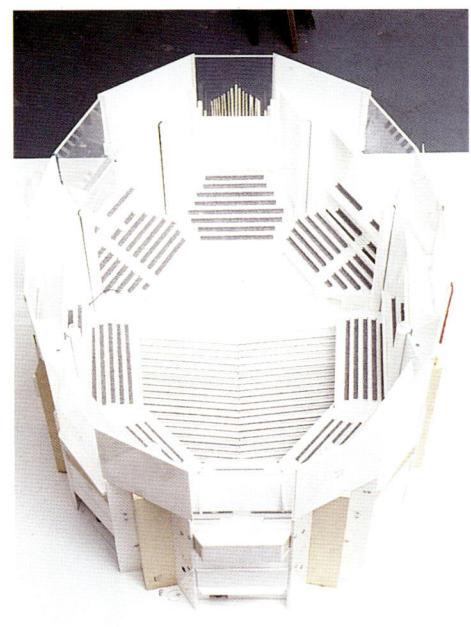

6

7

The natural contours of the site have been used to advantage. Car parks and the lower levels of the conference facilities will be built into the slope, and a series of terraces alongside the restaurants and cafes will enable the visitor to explore the gardens and surrounding natural woodland.

The first concert, scheduled for April 1997, will mark the opening of a major new concert venue in Europe.

场地的自然斜坡地形得到了利用。停车场和低层的会议室建在斜坡下方，挨着餐馆和咖啡厅则利用斜坡建造了一系列露台，使得游客能够观赏花园和周围自然林地的景色。

中心的首场音乐会是新举办的一场欧洲大型音乐节的开幕式，于1997年3月举行。

8

9

Istanbul Cultural and Congress Centre 211

Gro, Newtown

Appointed 1993
Newtown, Wales
Development Board for Rural Wales
Occupier: Control Techniques
2,900 square metres
Reinforced concrete and steel frame
Louvred cladding and curtain glazing

格罗大楼，新城

中标时间　1993
新城，威尔士
威尔士农村发展委员会
使用者：管理科技公司
2,900m²
钢筋混凝土，钢框架
散热覆层，玻璃幕墙

In 1993 the proposal to provide new research laboratories and offices for Control Techniques was approved by the Development Board for Rural Wales.

Planned for a site at Newtown, the scheme provides 2,900 square metres of space on two floors and is designed to accommodate both offices and research facilities. The curved building is linked to a circular glassed pavilion which houses a staff restaurant. Both offices and amenities have been designed to provide good levels of natural daylight and fine views over the surrounding countryside. A system of external screening protects windows from solar gain.

1

2

　　1993年，威尔士农村发展委员会批准了为管理科技公司建造新的研发实验室和办公室的提案。
　　场地位于新城，建筑分为两层，建筑面积2,900m²。建筑内包含办公室和研究设施。主建筑呈曲线形平面，和附属的一座圆形小楼相连接，小楼采用玻璃外装，内设员工餐厅。办公室和配套休息设施都采用了利于自然采光的设计，同时也可使人在楼内便利地眺望外部的乡村景色。外部的遮阳设施可以保护窗户免受过度日照的影响。

1　Block model showing the building and overall site layout
2　Model view showing the building in landscape context
3　Computer-generated typical floor plan
4　Computer-generated typical cross section

1　建筑及场地全景模型
2　建筑物及周围景观模型
3　计算机绘制的标准楼层平面图
4　计算机绘制的标准横剖面

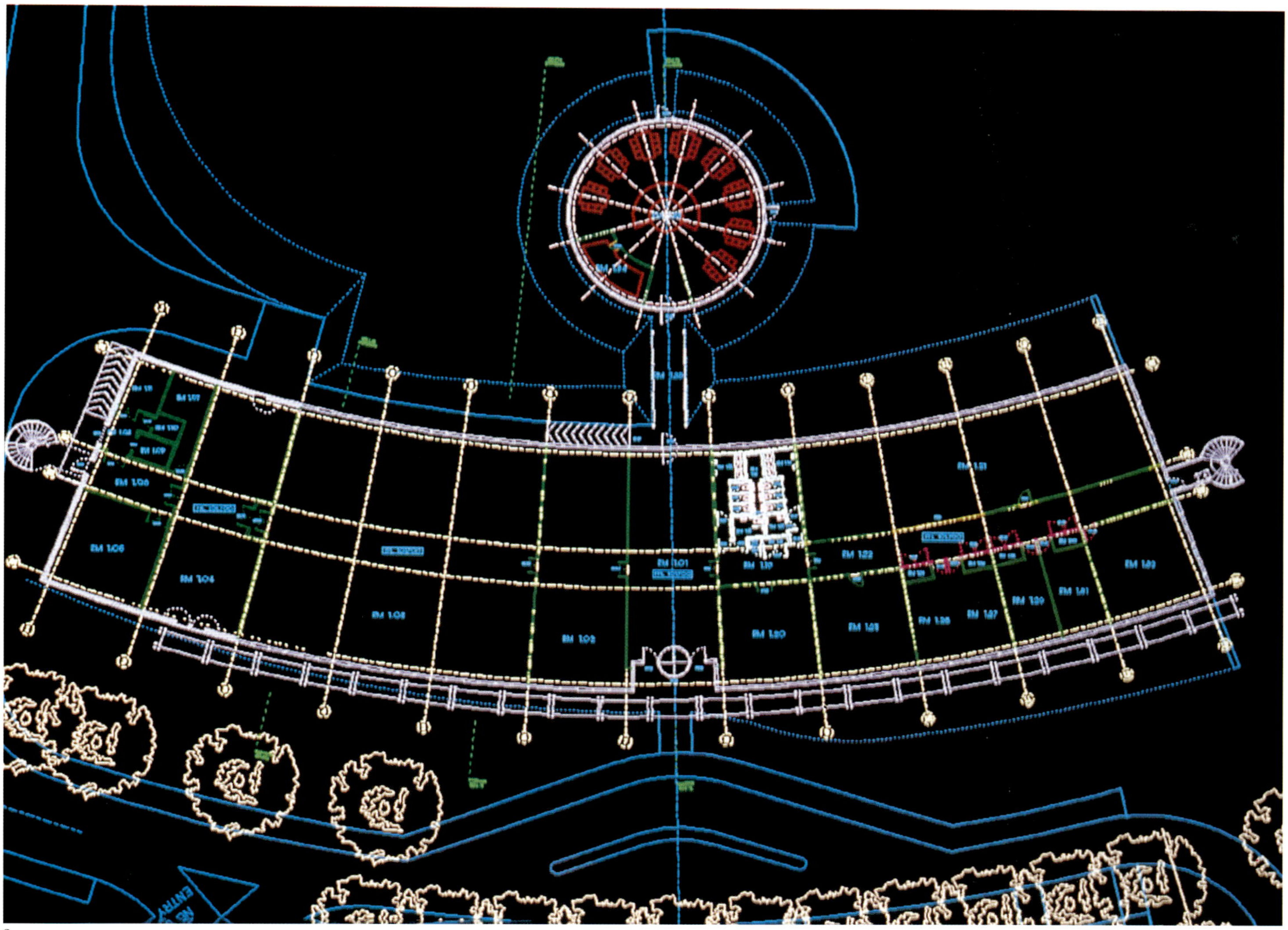

3

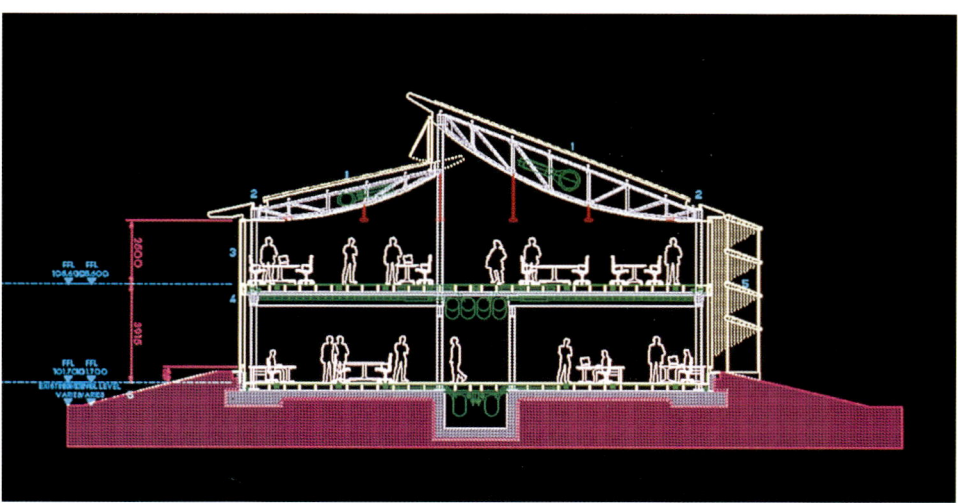
4

Manchester 2000 International Olympic Stadium

Design 1993
Manchester
AMEC Developments PLC
80,000 seating capacity
Galvanised steel
White fairfaced concrete, glass screen walls

曼彻斯特 2000 奥林匹克国际体育场

设计时间　　1993
曼彻斯特
AMEC 投资开发公司
座位容量：80,000
镀锌钢材
白色光面混凝土，玻璃幕墙

The design for the new International Olympic Stadium incorporates seating for 80,000 spectators and will create a landmark for the city.

The stadium is the focus of a master plan for the 48-hectare site—a development dedicated to sport with a range of facilities including a sports injuries hospital and sports industries business park. The site already incorporates a new velodrome for the British Cycling Federation.

The stadium, with a one kilometre circumference, has been designed to enhance spectator's sightlines and visual proximity to the sporting events.

Continued

新建的奥林匹克国际体育场设计座位容量为 80,000，将成为城市的标志建筑。

体育场位于占地 48hm² 的开发规划区域的中心位置，这块区域专门为体育活动开发建设，将建有一大批配套设施，包括一家运动损伤医院和相关的体育产业园区。开发区内还将建造一座供英国自行车运动联盟使用的室内赛车场。

体育场圆周长达 1km，设计者着重考虑了观众的视野范围，使观众观看比赛时可以得到更全面和更接近的视觉效果。

1

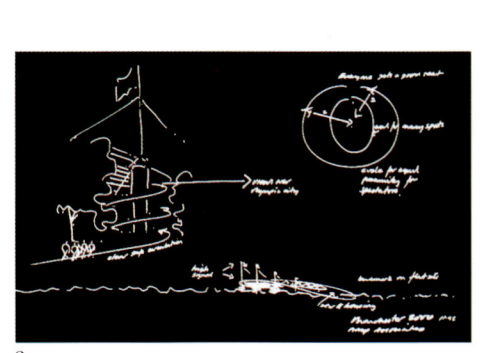
2

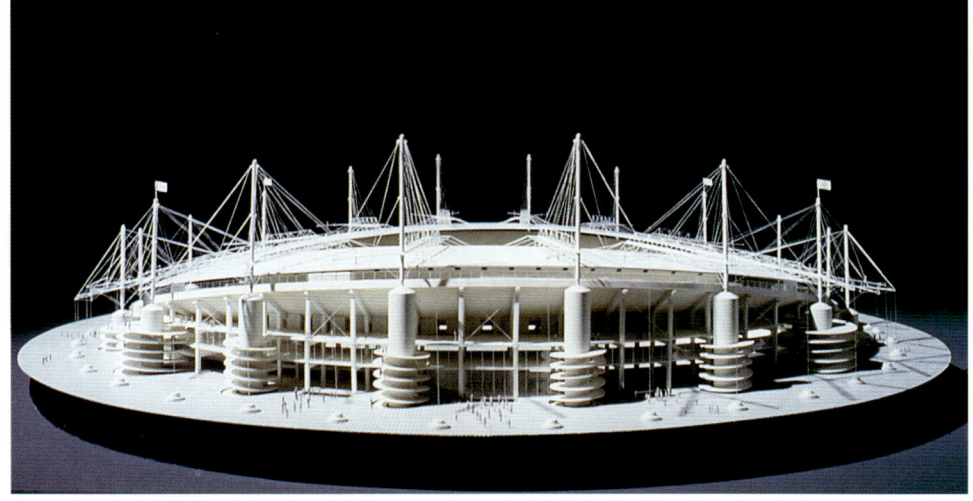
3

1　Site plan
2　Concept sketches
3　Model showing roofscape
4　Model as exhibited at The Royal Academy, London
5　Spiral access ramps integrated with vertical mast structure

1　场地平面图
2　设计概念草图
3　屋顶外景模型
4　在伦敦皇家艺术院展出的建筑模型
5　螺旋形的观众流通坡道与竖直的桅杆结构构成为一体

4

5

Manchester 2000 International Olympic Stadium 215

A circular seating plan combines with an oval arena to create elevations which sweep up from the north and south to contain four key levels on the east and west. The dramatic sweeping form moderates the bulk of the building. The high sides offer protection from prevailing winds and low sun angles while providing accommodation and a concentration of seats in the areas most favoured by spectators. The low sides allow sun onto the grassed arena and give a human scale to the structure.

The stadium is defined by a ring of structural towers. These towers support generous pedestrian circulation ramps which enhance both crowd comfort and safety. The towers also incorporate a mast and cable roof structure, giving a distinct visual order. They create a distinct image for the stadium and will be dramatically lit for night events.

6

7

体育场采用环形的看台和椭圆形的赛场，立面沿南北方向掠起，在东面和西面分四层。弯曲掠起的鲜明造型缓和了建筑物体量过大的压抑感。侧面中部较高，可以为观众集中的座位区域和主席台起到防风和减小日照角度的作用，两侧较矮的部分则可使赛场草坪得到直接的日照。

圆环形看台采用一系列塔形结构组成，塔楼中建有宽敞的环形人行坡道，可以更好地疏散拥挤的人群从而保证安全。塔楼还配套设计了桅杆和悬吊屋顶，使建筑外形十分独特，在夜间比赛的时候，还配以富有想像力的灯光，使建筑呈现灿烂而独特的外景。

6–7	Computer-generated view of stadium at night	6-7 计算机绘制的体育馆夜景图
8	Overall view of model	8 建筑模型全景图
9	Plan layout of Manchester stadium	9 曼彻斯特体育场平面布局图
10	Detail view of model showing pedestrian ramps for fast emergency exit	10 作为紧急快速出口的人行坡道详图

8

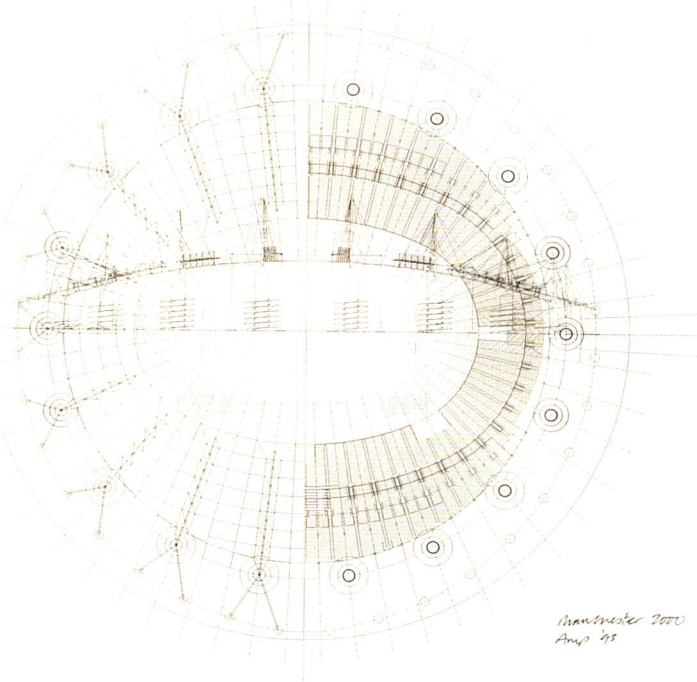
9

10

Manchester 2000 International Olympic Stadium 217

Johannesburg Athletics Stadium

Design/Completion 1993/1995
Johannesburg, South Africa
Johannesburg City Council
38,000-seat capacity
Steel and reinforced concrete
Light steel roof deck

约翰内斯堡田径运动场
设计/完成　　1993/1995
约翰内斯堡，南非
约翰内斯堡市政府
座位容量：38,000
钢材，钢筋混凝土
轻质钢屋盖

The site for this 38,000-seat stadium is adjacent to the existing Ellis Park rugby stadium. Both facilities will be used during the 1995 Rugby World Cup.

The stadium will serve as South Africa's premier athletics stadium for both local and international events. It will also function as a regular venue for soccer matches and open-air concerts.

The seating bowl which has been planned on the western side will house all the facilities for athletes, officials, competition management and the media. In addition the facility has been planned to provide function rooms for VIPs, hospitality boxes and a Sky Restaurant with commanding views over the city.

A delicate roof structure supported by steel masts and cables covers the seats on this side of the arena.

A second phase, which has been planned as an integral part of the design, would increase the capacity to 55,000 spectators by providing additional seating on the eastern side.

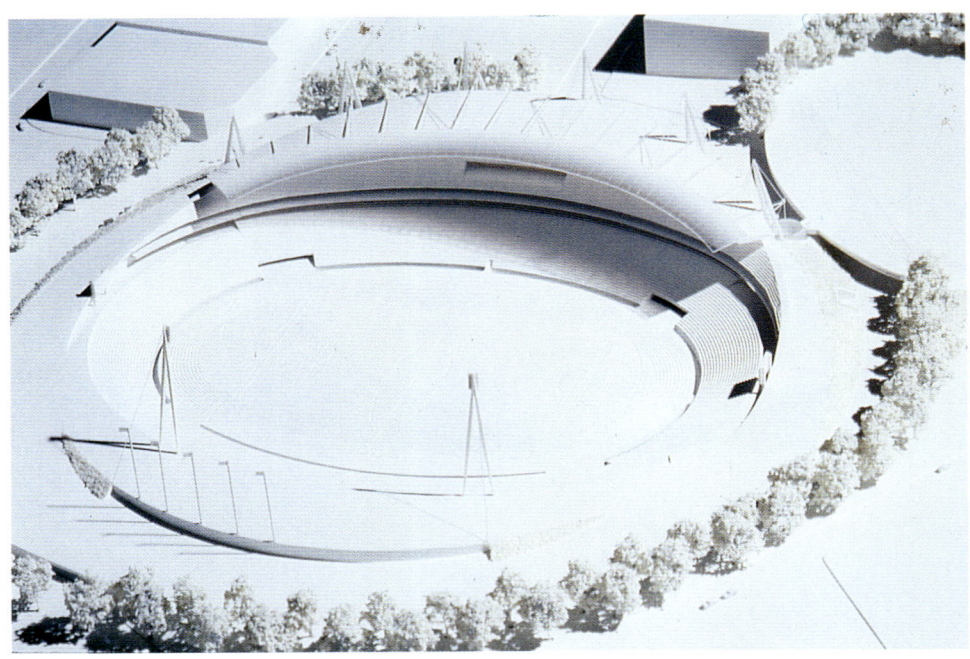
1

这座38,000座的体育场毗邻已有的埃利斯公园橄榄球场，这两座体育场都将在1995年世界杯橄榄球赛中使用。

这座体育场将是南非首要的田径运动场，用于举行国内和国外田径比赛，同时也可以用于举行正式的足球比赛和大型露天音乐会。

碗形的看台设计在西面，田径比赛的相关设施、官员、赛事管理机构和媒体也都安置在这里。另外，配套设施还包括贵宾休息室、医疗室和一间可居高俯瞰城市的空中餐厅。

看台采用轻质的屋顶进行遮挡，屋盖承重体系采用钢索和桅杆结构。

二期工程作为整个项目的一部分，计划将在赛场的东面增加观众席，使体育场的座位容量扩大至55,000。

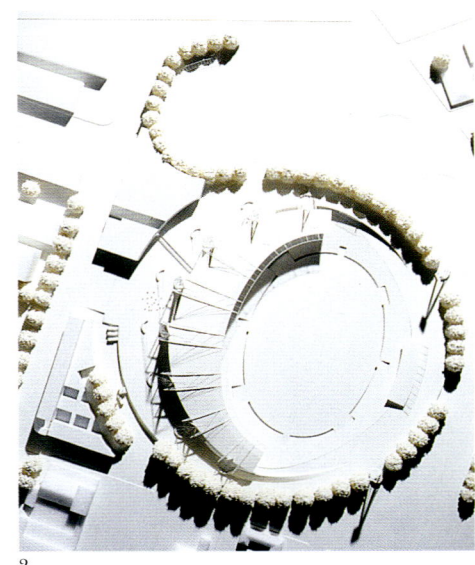
2

1 Model of cable net roof (developed later)
2 Competition-winning model
3 Conceptual sketch plan of stadium
4 Sketch of cable net roof

1 索网屋盖结构模型（后来得到采用）
2 在竞赛中获胜的模型
3 体育场设计概念草图
4 索网屋盖结构草图

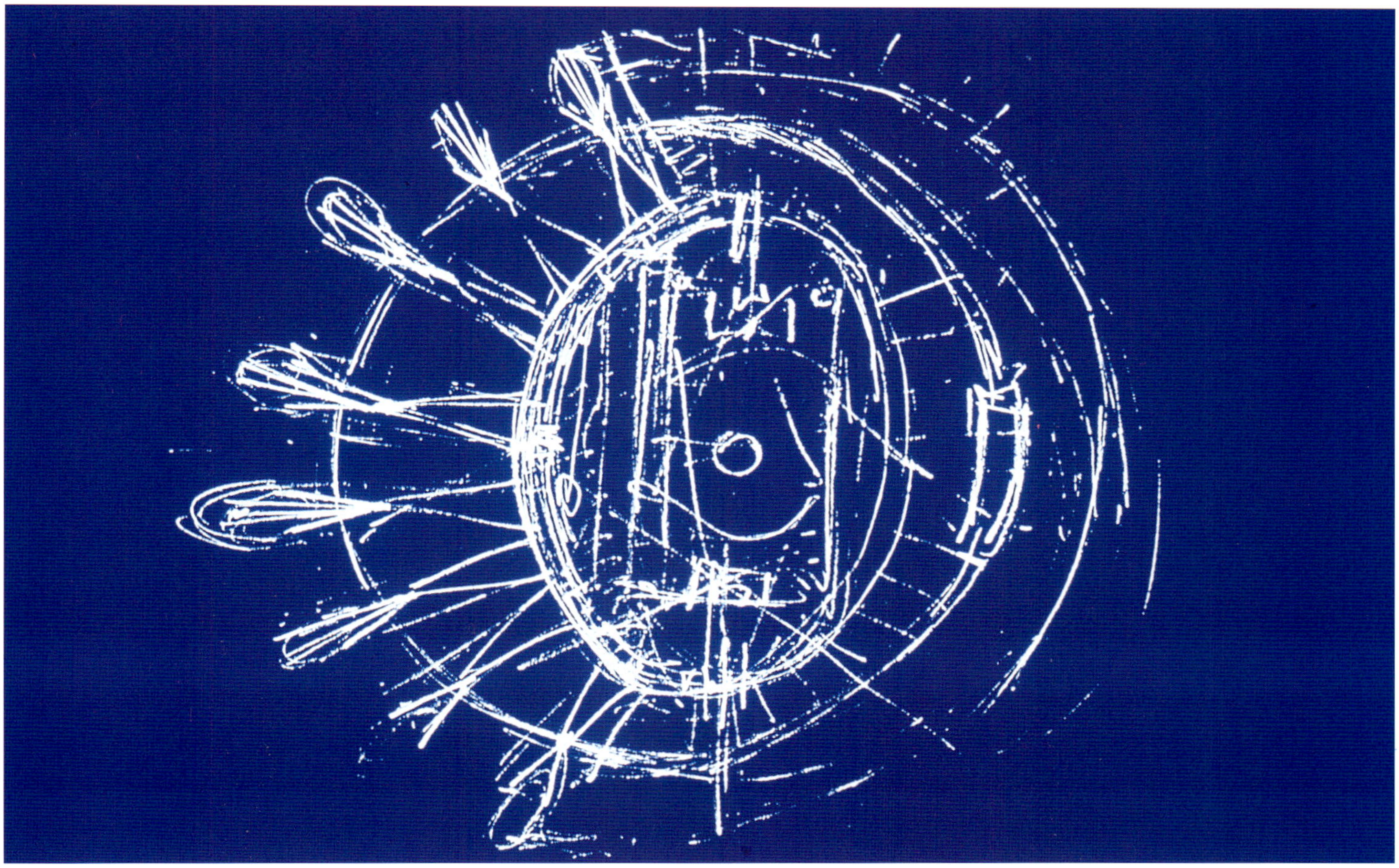

3

4

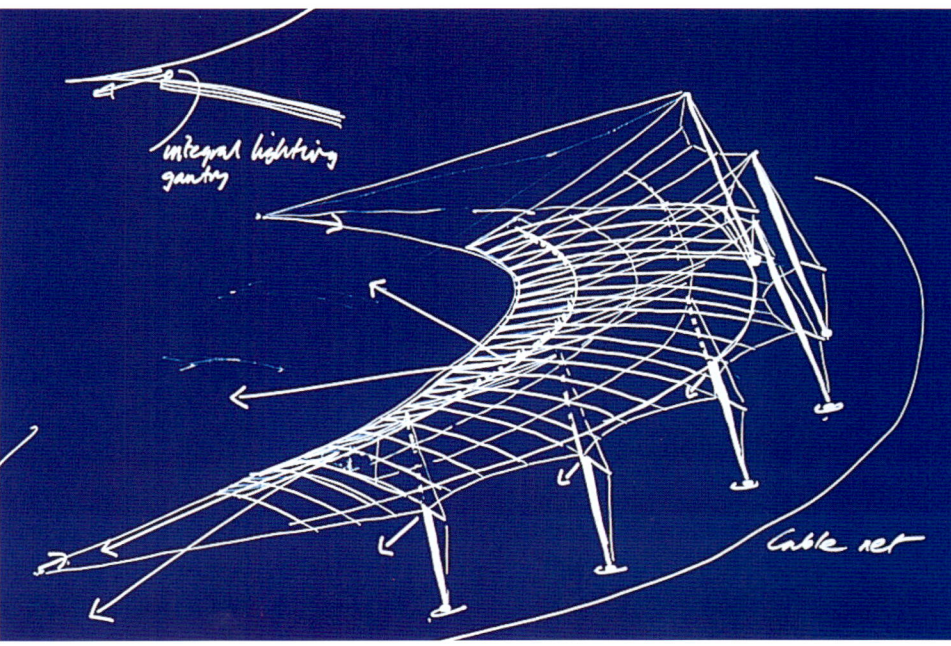

Johannesburg Athletics Stadium 219

Firm Profile

事务所简介

Biographies

个人简历

菲利普·道森爵士（SIR PHILIP DOWSON）

英帝国二等勋位爵士（CBE），文学硕士（MA），英国皇家艺术会会长（RA），建筑学专业毕业证书（AADip），皇家建筑师学会会员（RIBA），工业艺术与设计学会会员（FSIAD）
事务所创办合伙人，1993年后任事务所顾问

菲利普·道森曾在剑桥大学和建筑联合学院（the Architectural Association）学习建筑学，1953年毕业。1959年他加入奥夫·阿鲁普和合伙人事务所（Ove Arup & Partners），成为合伙人之一。他是1963年创办阿鲁普联合事务所的合作者之一，1969年成为高级合伙人。1971年至1975年任专业咨询委员会委员。他是皇家美术委员会委员，1979年后成为皇家艺术院会员。由于对建筑的杰出贡献，1969年他获得英帝国二等勋位爵士勋章并在1980年被授予爵位。1981年获得皇家建筑师学会金质奖章。

专业资质和会员资格：
剑桥大学文学硕士
英国皇家艺术会会长（1984）
建筑学专业毕业证书
皇家建筑师学会会员
工业艺术与设计学会会员
英国注册建筑师理事会注册执业资格
美国建筑师学会荣誉会员

戴维·登顿（DAVID DEIGHTON）

英国电机工程师学会会员（MIEE）

戴维·登顿1975年毕业于梅德威－梅德斯通（Medway & Maidstone）理工学院，大学专业为电机工程。1981年加入阿鲁普联合事务所，1990年成为主管。他开发出一套计算机软件用于建筑电气设备布局设计，负责过办公和商业楼电路设计及大型计算机设备的安装。

专业资质和会员资格：
工程学会委员会资质考试电机工程部分
项目管理执照
英国电机工程师学会会员

唐纳德·麦可·弗格森（DONALD MACKAY FERGUSON）

皇家建筑师学会会员

唐纳德·麦可·弗格森1964年毕业于布里斯托尔大学，大学专业为建筑学。1972年加入阿鲁普联合事务所，1984年成为主管。他设计过英国及海外的大型办公及休闲建筑，包括伊斯坦布尔文化和会议中心。1991年他成为阿鲁普联合事务所的董事之一。

专业资质和会员资格：
英国注册建筑师理事会注册执业资格
皇家建筑师学会会员

SIR PHILIP DOWSON
CBE MA PRA AADip RIBA FSIAD

Founder partner, consultant since 1993

Philip Dowson studied architecture at Cambridge and the Architectural Association, graduating in 1953. He joined Ove Arup & Partners and became an Associate in 1959. He was a founder partner of Arup Associates in 1963, and became a senior partner of the Ove Arup Partnership in 1969. He was a member of the Crafts Advisory Commission from 1971 to 1975, is a Member of the Royal Fine Art Commission and has been an Associate of the Royal Academy since 1979. For his services to architecture he was awarded a CBE in 1969 and was knighted in 1980. The following year he was the winner of the RIBA Gold Medal.

Professional qualifications and memberships
Master of Arts (Cantab)
President, Royal Academy, 1984
Diploma, Architectural Association
Member, Royal Institute of British Architects
Fellow of the Society of Industrial Artists and Designers
Registered Architect with the Architects' Registration Council of the United Kingdom
Honorary Fellow of the American Institute of Architects

DAVID DEIGHTON
MIEE

Joined Ove Arup & Partners 1992

David Deighton studied electrical engineering at Medway & Maidstone College of Technology, graduating in 1975. He joined Arup Associates in 1981 and became a director in 1990. He has developed a computer program to analyse electrical distribution in buildings, and has been responsible for the electrical design of offices, commercial buildings and large computer installations.

Professional qualifications and memberships
Council of Engineering Institution Part 2 Examination in Electrical Engineering
Diploma in Management Studies
Member, Institution of Electrical Engineers

DONALD MACKAY FERGUSON
RIBA

Don Ferguson studied architecture at Bristol University, graduating in 1964. He joined Arup Associates in 1972 and became a director in 1984. He has designed complex office buildings and leisure facilities both in the UK and overseas, including the culture and conference centre in Istanbul. He became a director of the Ove Arup Partnership in 1991.

Professional qualifications and memberships
Registered Architect with the Architects' Registration Council of the United Kingdom
Member, Royal Institute of British Architects

理查德·弗雷尔（RICHARD FREWER）
文学硕士，建筑学专业毕业证书（DipArch），皇家建筑师学会会员
1991年离开，1992年后任事务所顾问

理查德·弗雷尔曾在剑桥大学和建筑联合学院学习建筑学，1966年毕业。大学毕业后他就加入了阿鲁普联合事务所，1977年成为主管。他负责过教育研究、办公和住宅等多种类型建筑的设计。

专业资质和会员资格：
剑桥大学文学硕士
建筑学专业毕业证书
皇家建筑师学会会员
英国注册建筑师理事会注册执业资格

彼得·弗戈（PETER FOGGO）
建筑学学士，英帝国二等勋位爵士
1989年离开，1993年逝世

彼得·弗戈1957年毕业于利物浦大学，大学专业为建筑学。毕业后他开展私人设计业务，设计过学校建筑。1969年他成为阿鲁普联合事务所的合伙人之一，1984年成为高级合伙人。他负责过大学区开发规划及工业办公区开发项目。

专业资质和会员资格：
建筑学学士
英国注册建筑师理事会注册执业资格

阿拉斯泰尔·巴尔弗·古莱（ALASTAIR BALFOUR GOURLAY）
建筑学专业毕业证书，城市建筑艺术专业毕业证书（DipCD），皇家建筑师学会会员

阿拉斯泰尔·古莱1963年毕业于邓迪大学约旦斯通-邓肯学院，专业为建筑学。1965年他获得城市建筑艺术专业研究生毕业证书并加入了阿鲁普联合事务所，1984年成为主管。他负责过工业、办公和住宅区等建筑项目的开发，也从事过公共建筑的设计。

专业资质和会员资格：
建筑学专业毕业证书
城市建筑艺术专业毕业证书
皇家建筑师学会会员
英国注册建筑师理事会注册执业资格

RICHARD FREWER
MA AADip RIBA
Left 1991, Consultant since 1992

Richard Frewer studied architecture at Cambridge and at the Architectural Association, graduating in 1966. He joined Arup Associates in the same year and became a director in 1977. He has been responsible for the design of teaching and research buildings as well as office developments and residential projects.

Professional qualifications and memberships
Master of Arts (Cantab)
Diploma, Architectural Association
Member, Royal Institute of British Architects
Registered Architect with the Architects' Registration Council of the United Kingdom

PETER FOGGO
BArch CBE
Left 1989, died 1993

Peter Foggo studied architecture at Liverpool University. He graduated in 1957 and worked in private practice on the design of schools and buildings. He became a partner of Arup Associates in 1969 and a senior partner of the Ove Arup Partnership in 1984. He was responsible for university master planning and industrial and office developments.

Professional qualifications and memberships
Bachelor of Architecture
Registered Architect with the Architects' Registration Council of the United Kingdom

ALASTAIR BALFOUR GOURLAY
DipArch DipCD RIBA

Alastair Gourlay studied architecture at Duncan of Jordanstone College at Dundee University, graduating in 1963. Two years later he obtained a post-graduate diploma in civic design and joined Arup Associates. He became a director in 1984. He has been responsible for the design of industrial, office and residential developments as well as projects for public authorities.

Professional qualifications and memberships
Diploma Architecture
Diploma Civic Design
Member, Royal Institute of British Architects
Registered Architect with the Architects' Registration Council of the United Kingdom

Biographies Continued

约翰·黑尔（JOHN HARE）
文学硕士，皇家建筑师学会会员

约翰·黑尔曾在剑桥大学和哈佛大学设计研究院学习建筑学，1982年毕业。1984年加入阿鲁普联合事务所。他创建了计算机辅助设计系统，负责在公司引入新的设计系统并建立设计师培训体系。1989年他成为主管。

专业资质和会员资格：
剑桥大学文学硕士
皇家建筑师学会会员
英国注册建筑师理事会注册执业资格

罗纳德·霍布斯（RONALD HOBBS）
理学学士（BSc），英国土木工程师学会会员（MICE），英国结构工程师学会高级会员（FIStructE）
事务所创办合伙人，1989年后任事务所顾问

鲍勃·霍布斯1943年毕业于布里斯托尔大学，专业为土木工程。他曾在法伯勒的皇家空军基地工作。1948年他加入奥夫·阿鲁普和合伙人事务所，1955年成为合伙人之一，1961年成为高级合伙人。他是1963年创办阿鲁普联合事务所的合作者之一，1969年成为高级合伙人。他是英国建筑研究院工程与研究奖评审委员会委员，还是顾问工程师协会合同编审联合会代表。

专业资质和会员资格：
理学学士，工学学士（荣誉）
英国土木工程师学会会员，英国结构工程师协会高级会员
英国皇家建筑师学会荣誉高级会员

理查德·李（RICHARD LEE）
理学学士，英国土木工程师学会会员

理查德·李1972年在索尔福德大学获得土木工程专业学位。1976年他由奥夫·阿鲁普和合伙人事务所加入阿鲁普联合事务所，1990年成为主管。他曾从事一些大型商业开发区的设计和规划。

专业资质和会员资格：
理学学士（荣誉），土木工程专业学位
英国土木工程师学会会员

JOHN HARE
MA RIBA
Left 1992

John Hare studied architecture at Cambridge University and Harvard Graduate School of Design, graduating in 1982. He joined Arup Associates in 1984 to establish the Computer Aided Design system and was responsible for introducing new systems and setting up the training program for designers in all the professions. He became a director in 1989.

Professional qualifications and memberships
Master of Arts (Cantab)
Member, Royal Institute of British Architects
Registered Architect with the Architects' Registration Council of the United Kingdom

RONALD HOBBS
BSc MICE FIStructE
Founder partner, Consultant since 1989

Bob Hobbs graduated in civil engineering from Bristol University in 1943 and worked with the Royal Aircraft Establishment at Farnborough. He joined Ove Arup & Partners in 1948 and became an associate in 1955. In 1961 he became a senior partner of Ove Arup & Partners and in 1963 was a founder partner of Arup Associates. In 1969 he became a senior partner of the Ove Arup Partnership. He is a member of the Projects and Research Awards Panel for ARCUK and represents the Association of Consulting Engineers on the Joint Contracts Tribunal.

Professional qualifications and memberships
Bachelor of Science, Engineering (Honours)
Member, Institute of Civil Engineers
Fellow, Institute of Structural Engineers
Honorary Fellow of the Royal Institute of British Architects

RICHARD LEE
BSc MICE

Richard Lee obtained his degree in civil engineering at Salford University in 1972. In 1976 he joined Arup Associates from Ove Arup & Partners, as a civil and structural engineer. He became a director in 1990 and has been involved in the design and co-ordination of several major commercial developments.

Professional qualifications and memberships
Bachelor of Science (Honours), Civil Engineering
Member, Institute of Civil Engineers

迈克尔·艾伦·洛厄尔（MICHAEL ALAN LOWE）
建筑学学士（BArch），建筑学和城市规划硕士（March + UD），皇家建筑师学会会员，美国城市规划学会会员（AIP）

  迈克尔·洛厄尔曾在南非开普敦大学、美国圣路易斯大学学习建筑学，于1966年获得建筑学和城市规划硕士学位。他曾在南非和北美工作并积累了一定经验。1978年他进入阿鲁普联合事务所，主要负责英国和欧洲的大型城市规划项目，包括科技和商业园区的开发项目。1988年他成为主管。

  专业资质和会员资格：
  建筑学学士
  建筑学和城市规划硕士
  皇家建筑师学会会员
  美国城市规划学会会员
  英国注册建筑师理事会注册执业资格

约翰·帕克（JOHN PARK）
皇家特许测量师学会预备会员（ARICS）
1989年离开

  约翰·帕克曾在布里克斯顿建筑学校和房地产管理学院学习，并获得特许测量师资格。1967年他进入阿鲁普联合事务所，1988年成为主管。他参与了项目控制体系和管理合同的编制开发工作。

  专业资质和会员资格：
  皇家特许测量师学会预备会员

特伦斯·拉格特（TERENCE RAGGETT）
英国结构工程师学会会员

  特伦斯·拉格特1969年毕业于吉尔福德建工学院，专业为工程和建筑技术。1974年他进入阿鲁普联合事务所，1984年成为主管。他参与了许多工业和办公建筑的设计，还参与了沙特阿拉伯的商业娱乐区的开发。

  专业资质和会员资格：
  英国结构工程师学会会员
  英国国家高级学位证书

约翰斯顿·里德尔（JOHNSTON RIDDELL）
皇家特许测量师学会预备会员
1993年离开

  约翰斯顿·里德尔1960年毕业于荷洛特·瓦特大学，专业为测量学。毕业后他从事私营业务。1974年他进入阿鲁普联合事务所，1981年成为主管。他曾担任过大型商业区和大学城项目的高级工程测量师。

  专业资质和会员资格：
  皇家特许测量师学会预备会员

MICHAEL ALAN LOWE
BArch MArch+UD RIBA AIP

Michael Lowe studied architecture in Cape Town, South Africa and St Louis, USA, and obtained his master's degree in architecture and urban design in 1966. He has gained experience by working in Africa and North America. He joined Arup Associates in 1978 and has primarily been responsible for large urban planning projects in the UK and Europe, including the development of science and business parks. He became a director in 1988.

Professional qualifications and memberships
Bachelor of Architecture
Master of Architecture and Urban Design
Member, Royal Institute of British Architects
Member, American Institute of Planners
Registered Architect with the Architects' Registration Council of the United Kingdom

JOHN PARK
ARICS

Left 1989

John Park studied at Brixton School of Building and College of Estate Management and is a chartered surveyor. He joined Arup Associates in 1967 and became a director in 1988. While at Arup Associates he was involved in the development of the Project Control System and pioneered Management Contracting.

Professional qualifications
Associate, Royal Institute of Chartered Surveyors

TERENCE RAGGETT
MIStructE

Terry Raggett studied engineering and building technology at Guildford College of Building, graduating in 1969. He joined Arup Associates in 1974 and became a director in 1984. He has been involved in the structural design of industrial and office buildings as well as commercial and leisure developments in Saudi Arabia.

Professional qualifications and memberships
Member, Institute of Structural Engineers
Higher National Diploma

JOHNSTON RIDDELL
ARICS

Left 1993

Jo Riddell studied surveying at Heriot Watt University, graduating in 1960. He worked in private practice until he joined Arup Associates in 1967, becoming a director in 1981. He has been the senior project quantity surveyor on major commercial and university projects.

Professional qualifications and memberships
Associate, Royal Institute of Chartered Surveyors

Biographies Continued

彼得·斯基德（PETER SKEAD）
理学学士，南非土木工程师学会会员（SAICE）

彼得·斯基德1957年毕业于南非开普敦大学，专业为土木工程。1958年他加入奥夫·阿鲁普和合伙人事务所，1963年成为阿鲁普联合事务所的创办合作者之一。他曾担任许多重要项目的高级工程师，1989年他成为主管。

专业资质和会员资格：
理学学士，土木工程专业学位
南非土木工程师学会会员

德里克·苏格登（DEREK SUGDEN）
英国土木工程师学会会员，英国结构工程师学会会员，焊接学会会员（MWeldI），声学工程师学会会员（MIOA）
创办合伙人之一，1987年退休

德里克·苏格登曾在威斯敏斯特理工学院学习，并获得土木和结构工程师资格。他是1963年阿鲁普联合事务所的创办合作者之一。他在工业建筑钢结构设计方面有独到的经验。1980年他成为阿鲁普联合声学工程设计事务所的创办人之一。他是南岸理工学院理事会的成员，还是伦敦大学学院巴特利特建筑规划分校的访问教授。

专业资质和会员资格：
英国土木工程师学会会员
英国结构工程师学会会员
焊接学会会员
声学工程师学会会员

戴维·爱德华·托马斯（DAVID EDWARD THOMAS）
建筑学学士（荣誉），皇家建筑师学会会员

戴维·托马斯1957年毕业于利物浦大学，专业是建筑学。毕业后他曾从事过医院建筑的设计。1964年他进入阿鲁普联合事务所，1977年成为主管。他曾负责过大型办公和工业建筑的开发规划。

专业资质和会员资格：
建筑学学士（荣誉）
英国皇家建筑师学会特许会员
英国注册建筑师理事会注册执业资格

PETER SKEAD
BSc SAICE

Peter Skead studied civil and structural engineering at the University of Cape Town, South Africa, graduating in 1957. The following year he joined Ove Arup & Partners and became a founder member of Arup Associates when it was created in 1963. He has been senior project engineer on many important projects and became a director in 1989.

Professional qualifications and memberships
Bachelor of Science, Civil Engineering
Member, South African Institution of Civil Engineers

DEREK SUGDEN
MICE MIStructE MWeldI MIOA

Founder partner, retired 1987

Derek Sugden qualified as a civil and structural engineer at Westminster Technical College and worked for contractors and consulting engineers. He was a founder partner of Arup Associates in 1963. He has particular experience in the field of structural steelwork design for industrial buildings. In 1980 he was a founder member of Arup Acoustics. He is a Member of the Council of Governors, Polytechnic of the South Bank and a Visiting Professor at Bartlett School of Architecture and Planning, University College London.

Professional qualifications and memberships
Member, Institute of Civil Engineers
Member, Institute of Structural Engineers
Member of the Welding Institute
Member of the Institute of Acousticians

DAVID EDWARD THOMAS
BArch (Hons) RIBA

David Thomas studied architecture at Liverpool University and graduated in 1957, after which he worked on the design of hospitals. He joined Arup Associates in 1964 and became a director in 1977. He has been responsible for the development planning of major offices and industrial buildings.

Professional qualifications and memberships
Bachelor of Architecture (Honours)
Chartered Member, Royal Institute of British Architects
Registered Architect with the Architects' Registration Council of the United Kingdom

布鲁斯·维克斯（BRUCE VICKERS）
皇家特许测量师学会预备会员
1989年离开

　　布鲁斯·维克斯曾在西南艾塞克斯理工学院学习，1964年获特许测量师资格，1977年成为阿鲁普联合事务所的合伙人之一。他负责开发了实现有效成本控制和报表的相关程序。

　　专业资质和会员资格：
　　皇家特许测量师学会预备会员

彼得·沃伯顿（PETER WARBURTON）
建筑设备工程师特许协会高级会员（FCIBSE）

　　彼得·沃伯顿1966年获得从业资格认证，在此之前一直以学徒身份为国际建筑设备承包商工作。1969他进入阿鲁普联合事务所，1977年成为主管。他深入参与了建筑环境控制设计的计算机软件的开发。他是建筑设备工程师特许协会计算机应用研究小组的主席。

　　专业资质和会员资格：
　　暖通和空调工程执照
　　英国国家高级学位证书（荣誉）
　　工程管理专业毕业证书（荣誉）
　　建筑设备工程师特许学会高级会员
　　美国暖通空调工程师学会会员

查尔斯·瓦默（CHARLES WYMER）
文学士，英国土木工程师学会会员，英国结构工程师学会会员

　　查尔斯·瓦默1961年毕业于牛津大学，专业为工程学。1963年他进入ARUP事务所，1977年成为主管。在事务所他负责大学城的开发规划以及教学研究建筑的设计，他也参与过工业、计算机和办公项目的设计。

　　专业资质和会员资格：
　　文学士
　　英国土木工程师学会会员
　　英国结构工程师学会会员

BRUCE VICKERS
ARICS

Left 1989

Bruce Vickers studied at the South West Essex Technical College and became a chartered surveyor in 1964. In 1977 he became a partner of Arup Associates. He was responsible for evolving programs for effective cost control and reporting techniques.

Professional qualifications
Associate, Royal Institute of Chartered Surveyors

PETER WARBURTON
FCIBSE

Peter Warburton served an apprenticeship with international services contractors before obtaining degree qualifications in 1966. He joined Arup Associates in 1969 and became a director in 1977. He has been deeply involved in the development of computer programs for environmental control design. He is Chairman of the Computer Applications Panel of the Chartered Institution of Building Services.

Professional qualifications and memberships
Diploma in Heating, Ventilation, Air Conditioning and Fan Engineering
Higher National Certificate (Distinction)
Diploma in Engineering Management (Honours)
Fellow, Chartered Institution of Building Services Engineers
Member, American Society of Heating, Ventilation and Refrigeration Engineers

CHARLES WYMER
BA MICE MIStructE

Retired 1992, died 1993

Charles Wymer studied engineering at Oxford University and graduated in 1961. Two years later he joined Arup Associates and became a director in 1977. At Arup Associates he was responsible for university development plans and for the structural design of teaching and research buildings as well as industrial, computer and office projects.

Professional qualifications
Bachelor of Arts
Member, Institute of Civil Engineers
Member, Institute of Structural Engineers

Chronological List of Buildings & Projects   建筑及项目年表

*Indicates work featured in this book
(see Selected and Current Works).

工厂
1955
亨普斯特德，赫特福德郡
化工建筑制品公司

环氧树脂工厂
1958
杜克斯津，剑桥郡
CIBA 有限公司

研究实验室
1959
杜克斯津，剑桥郡
CIBA 有限公司

化学工厂
1961
敦桥，肯特郡
史密斯·克兰和弗伦奇实验室

邮局
1961
海角，加纳
邮政电信局

工厂，实验室和办公楼
1961
韦尔温花园城，赫特福德郡
史密斯·克兰和弗伦奇实验室

维护站
1962
杜克斯津，剑桥郡
CIBA 有限公司

约克－什普利有限公司工厂
1962
巴兹尔登，艾塞克斯郡
约克－博格瓦那有限公司

粘合剂工厂
1962
斯塔福德，斯塔福德郡
埃沃德有限公司

办公楼
1963
韦尔温花园城，赫特福德郡
史密斯·克兰和弗伦奇实验室有限公司

工厂
1963
伦敦北一区
弗罗林有限公司

清漆工厂
1963
斯塔福德，斯塔福德郡
埃沃德有限公司

多功能大楼
1964
杜克斯津，剑桥郡
CIBA 有限公司

工业研究实验室
1964
霍沙姆，萨里郡
CIBA 有限公司

圣体学院，雷克汉普顿楼
1964
剑桥，剑桥郡
剑桥大学

工厂
1964
阿什津，肯特郡
沃尔特－琼斯有限公司

多层停车场
1964
赫尔，亨伯赛德郡
L. 比尔和后裔有限公司

皇家独立公寓
1964
布拉克内尔，伯克郡
布拉克内尔开发公司

福尔马林工厂
1965
杜克斯津，剑桥郡
CIBA 有限公司

游泳池
1965
泰晤士河畔的沃尔顿，萨里郡
泰晤士河畔的沃尔顿市政府

*伯明翰大学采矿和冶金学院
1966
伯明翰
伯明翰大学

维克产品工厂
1966
斯塔福德，斯塔福德郡
埃沃德有限公司

造纸厂
1966
杰伯，尼日利亚
Coutinho Caro Hamburg

电话局
1966
阿克拉北区，加纳
邮政电信局

斯科特斯通楼
1966
南石板渡，洛锡安区
奥雅纳合伙人事务所

阿登布鲁克斯区开发规划
1966
剑桥，剑桥郡
剑桥大学

拉伯运大学总规划
1966
拉伯运，勒斯特郡
拉伯运大学

沃尔夫森楼，萨默维尔学院
1967
牛津，牛津郡
牛津大学

* 莫尔丁斯音乐厅
1967
斯耐普，阿尔德柏夫，萨福克郡
阿尔德柏夫音乐节

福莱楼，萨默维尔学院
1967
牛津，牛津郡
萨默维尔学院，牛津大学

伯尔顿楼，三一公寓
1968
剑桥，剑桥郡
剑桥大学

仓库和书店
1968
哈蒙德斯沃斯，大伦敦地区
企鹅图书有限公司

住宅楼
1968
拉伯运，勒斯特郡
拉伯运大学

土木系
1968
拉伯运，勒斯特郡
拉伯运大学

* IBM（英国）计算机制造装配中心
1969
哈万特，汉普郡
IBM（英国）有限公司

私人住宅
1969
德拉克斯大道，温布尔登，伦敦西南20区
J·桑兹先生

* IBM 公司大楼，约翰内斯堡
1969
约翰内斯堡，南非
索桑斯房地产有限公司

人文与社会社学系，阿特伯勒塔楼
1970
勒斯特，勒斯特郡
勒斯特大学

化学工程系
1970
拉伯运，勒斯特郡
拉伯运大学

机械工程系
1970
拉伯运，勒斯特郡
拉伯运大学

电子显微镜大楼
1970
牛津，牛津郡
牛津大学

雄狮广场城市中心开发规划
1970
剑桥，剑桥郡
剑桥城

文学和商学院，穆尔海德大楼
1970
伯明翰
伯明翰大学

* IBM 公司北港工程
1970
北港，朴次茅斯，汉普郡
IBM（英国）有限公司

核物理大楼，牛津，三期工程
1971
牛津，牛津郡
牛津大学

体育馆
1971
吉尔福德，萨里郡
萨里大学

Chronological List of Buildings & Projects 229

Chronological List of Buildings & Projects Continued

新博物馆
1971
剑桥，剑桥郡
剑桥大学

微生物学系
1971
伯明翰
伯明翰大学

＊地平线项目，约翰·普莱耶和后裔的工厂
1971
诺丁汉
帝国烟草集团有限公司

电子显微镜大楼
1971
伯明翰
伯明翰大学

谢菲尔德大学开发规划
1971
谢菲尔德
谢菲尔德大学

印刷厂及办公楼
1971
牛津，牛津郡
牛津邮政时报

沃恩大楼
1972
牛津，牛津郡
萨默维尔学院，牛津大学

办公楼
1972
哈蒙德斯沃斯，大伦敦地区
企鹅图书有限公司

休息所
1972
考斯，威特岛
皇家游艇编队

办公楼和仓库
1972
伯利圣埃德蒙斯，萨福克郡
朗曼企鹅有限公司

运动中心
1972
安普福思，北约克郡
安普福思学院

威其菲尔德楼，三一公寓
1972
剑桥，剑桥郡
三一公寓，剑桥

中心办公楼
1973
彼得伯勒，剑桥郡
英国糖业公司

多层停车场
1973
雄狮广场城市中心开发规划
剑桥，剑桥郡
剑桥城

皇家海军基地厚钢板制造厂
1973
朴次茅斯，汉普郡
环境部

办公楼
1973
彼得伯勒，剑桥郡
英国糖业公司

西南风项目发展规划
1973
毛里求斯西南部
西南旅游开发有限公司

音乐学院
1973
诺里奇，东英吉利亚
东英吉利亚大学

住宅区
1974
斯塔沃顿大道，牛津，牛津郡
牛津大学学院

阿斯顿大学开发规划
1974
伯明翰
阿斯顿大学

教学楼
1974
吉尔福德，萨里郡
萨里大学

物理系、电机系和化工系
1974
拉伯运，勒斯特郡
拉伯运大学

图书馆和商场
1975
雄狮广场城市中心开发规划
剑桥，剑桥郡
剑桥城

学校建筑
1975
安普福思，北约克郡
安普福思大学

*亨利·伍德音乐厅
1975
南瓦克，伦敦东南1区
南瓦克音乐厅有限公司

皇家剧院修复工程
1975
格拉斯哥
苏格兰歌剧院

计算机中心
1975
谢菲尔德，南约克郡
谢菲尔德大学

仓库和住宅
1976
沃里克大道，伦敦西南5区
皇家肯斯顿和切尔西自治区

*布什·莱恩大楼
1976
坎农大街，伦敦东部中央4区
卡法格房屋发展有限公司

小卖部
1976
彼得伯勒，剑桥郡
英国糖业公司

*盖特威1号楼
1976
贝辛斯托克，汉普郡
维金斯蒂普（英国）公司

*杜鲁门有限公司总部
1976
砖巷，伦敦东1区
杜鲁门有限公司

*托马斯·怀特爵士大楼，牛津圣约翰学院
1976
牛津，牛津郡
圣约翰学院

冶金系
1976
牛津，牛津郡
牛津大学

皇家海事附属维护检修和供给中心
1977
朴次茅斯，汉普郡
环境部

计算机中心，三期工程
1977
科莎姆，汉普郡
IBM（英国）有限公司

皇家海军基地码头大楼
1977
朴次茅斯，汉普郡
环境部

公共汽车站及办公楼
1977
北安普敦，北安普敦郡
北安普敦市政府

伦敦大剧院
1977
伦敦中央西2区
英国国家歌剧院

*伦敦劳埃德保险公司总部
1978
查莎姆，肯特郡
劳埃德保险公司

*英国中央电力局西南区总部
1978
北德明斯特高地，布里斯托尔
英国中央电力局

皮尔斯－布里顿音乐学校
1978
斯耐普，萨福克郡
斯耐普－莫尔丁斯基金会

住宅
1978
埃弗，白金汉郡
P·赫德先生

罐头厂
1978
麦尔顿－莫伯诺，勒斯特郡
帕迪格雷宠物食品有限公司

化工厂
1978
牛顿－埃克里弗，达累姆郡
英国塑料工业有限公司

Chronological List of Buildings & Projects 231

Chronological List of Buildings & Projects Continued

产品仓库
1978
麦尔顿－莫伯诺，勒斯特郡
帕迪格雷宠物食品有限公司

歌剧院改造
1979
布克斯顿，德贝郡
布克斯顿歌剧院有限公司

雄狮广场地方法院
1979
剑桥，剑桥郡
剑桥城

游泳池
1980
伊顿，伯克郡
伊顿学院

*特房伯有限公司工厂
1980
柯彻斯特，艾塞克斯郡
特房伯有限公司

系统装配车间及办公室
1981
雷丁，伯克郡
数码设备有限公司

贝德福德学校
1981
贝德福德，贝德福德郡
贝德福德慈善机构

写字楼及商业中心
1981
达曼，沙特阿拉伯
沙特阿拉伯 GOSI 公司

津巴布韦房屋改造
1981
斯特兰德大道，伦敦中央西 2 区
建设部

市中心巴伯舍夫住宅区及商业区
1981
巴格达，伊拉克
阿玛纳特和阿斯玛公司

利物大厦
1982
利物浦
利物浦市政署

*巴伯夫区自治会办公楼
1982
哈德雷夫，萨福克郡
巴伯夫区自治会

阿莫干伯停车场及商业中心
1982
沙特阿拉伯
塔拉商业机构

希拉特－阿哈玛德停车场及商业中心
1982
沙特阿拉伯
塔拉商业机构

*盖特威 2 号楼
1982
贝辛斯托克，汉普郡
维金斯蒂普（英国）公司

露天游泳池
1983
伊顿，伯克郡
伊顿学院

外交使馆区警察局
1983
利雅得，沙特阿拉伯
外交事务署，沙特阿拉伯

*布莱克里夫大楼
1983
法伯勒，汉普郡
皇家养老金信托集团有限公司

科学实验室
1984
伊顿，伯克郡
伊顿学院

*国际园艺节展厅
1984
利物浦
默西塞德郡开发公司

港区轻轨铁铁路设计咨询
1984
伦敦
伦敦港区开发公司

*外交使馆区运动俱乐部
1985
利雅得，沙特阿拉伯
外交事务署，沙特阿拉伯

伦敦诺丁山公交车站
1985
伦敦中央东 2 区
伦敦交通局

伦敦水湾公交车站
1985
伦敦西 11 区
伦敦交通局

贝德福德学校
1985
贝德福德，贝德福德郡
贝德福德慈善团体

伦敦劳埃德出版社
1986
柯彻斯特，艾塞克斯郡
伦敦劳埃德公司

*福布斯·梅隆图书馆，克莱尔学院，剑桥
1986
剑桥，剑桥郡
克莱尔学院，剑桥

*斯托克利园区
1986
希思罗，优克斯桥，米德尔塞克斯郡
斯托克利园财团有限公司

斯克普大楼
1987
贝辛斯托克，汉普郡
古尔德公司

伦敦河堤公交车站
1987
伦敦中央西 2 区
伦敦交通局

*主祷广场
1987
伦敦东部中央 1 区
帕特诺斯特投资开发公司

皮货大楼
1988
伦敦中央西 2 区
高级皮货公司

沙特阿拉伯国家保护区
1988
迪拉伯，沙特阿拉伯
科昂纳姆企业

*哈斯伯罗－布拉德利（英国）有限公司总部
1988
斯托克利园区，希思罗，优克斯桥，米德尔塞克斯郡
哈斯伯罗－布拉德利（英国）有限公司

米尔顿－布莱德利设备公司 B1 楼
1988
斯托克利园区，希思罗，优克斯桥，米德尔塞克斯郡
米尔顿－布莱德利设备公司

*布鲁德门
1988
伦敦中央东 2 区
洛斯豪夫·斯坦诺普投资开发公司

*法人－通用保险公司大楼
1988
金斯伍德，萨里郡
法人－通用保险有限公司

*芬斯伯里大道 1 号
1988
伦敦东部中央 2 区
洛斯豪夫·格雷可特房地产开发有限公司

帝国战争博物馆一期工程
1989
朗伯斯区，伦敦东南 1 区
公共资产机构

*斯托克利园区会所
1989
斯托克利园区，优克斯桥，米德尔塞克斯郡
斯托克利园财团有限公司

斯托克利园区 A3.2 楼
1989
斯托克利园区，优克斯桥，米德尔塞克斯郡
信托证券资产管理有限公司

温布尔登桥大楼
1990
温布尔登，伦敦西南 19 区
伦敦默顿自治区

斯托克利园区 A3 楼
1990
斯托克利园区，优克斯桥，米德尔塞克斯郡
信托证券资产管理有限公司

斯托克利园区 B5 楼
1990
斯托克利园区，优克斯桥，米德尔塞克斯郡
斯托克利园财团有限公司

*温特沃斯高尔夫球俱乐部
1990
温特沃斯，萨里郡
切尔西菲尔德公司/温特沃斯资产机构

Chronological List of Buildings & Projects Continued

曼彻斯特机场停车场
1990
曼彻斯特
阿尔弗雷德-麦克阿尔卑斯有限公司

威尔本-维特项目
1990
奥利,法国
切尔西菲尔德公司

滨益公园
1990
北海道,日本
东京 M.I.O 有限公司

* 苏塞克斯正面看台,古德伍德赛马场
1990
古德伍德,苏塞克斯郡
古德伍德赛马场有限公司

* 白金汉宫大道 123 号
1990
伦敦东南一区
格雷科特房地产开发有限公司

* 皇家保险大楼
1991
彼得伯勒,剑桥郡
皇家人寿保险资产有限公司

* 劳埃德银行股份公司办公大楼
1991
卡能斯滩地,布里斯托尔
劳埃德银行股份公司

* 科坡斯恩酒店
1991
泰恩河畔的纽卡斯尔
科坡斯恩酒店(纽卡斯尔)有限公司

沃特令大楼
1991
伦敦中央东 4 区
斯坦荷普开发公司

因玛萨特广场
1991
伦敦西北 1 区
因玛萨特 OAP 公司

联合大农场
1991
伯恩机场,剑桥郡
斯坦荷普开发公司

植物园
1991
伦敦中央东 3 区
英国土地投资有限公司

政府办公大楼
1991
玛莎姆大街 2 号,伦敦西南 1 区
环境部

科技园
1991
布达诺瓦,匈牙利
电信工业集团公司

办公楼及商场
1991
牛津大街,伦敦西 1 区
卡加尔外贸公司

伯林顿园区
1991
普利茅斯,德文郡
R·堪迪和后裔(农场)有限公司

牛津火车站
1991
牛津,牛津郡
圣·凯瑟琳有限公司

XXth 园区
布达佩斯,匈牙利
电信工业集团公司

* 朱比利地铁线附属服务控制中心
1991
尼斯登,伦敦西北 10 区
伦敦地铁有限公司/朱比利地铁线附属设施

* 霍莎姆园区,盖莎根
1991
盖莎根,德国
霍莎姆投资开发公司

* 伯格塞尔露天赛场
1991
因斯布鲁克,奥地利
设计竞赛参赛作品

柯克大学校园
1991
伊斯坦布尔,土耳其
柯克基金会

特莱登广场
1992
伦敦西北 1 区
英国土地投资有限公司

*古维尔和卡尔斯学院，剑桥
1992
剑桥，剑桥郡
古维尔－卡尔斯学院

伦敦卢顿机场
1992
卢顿，贝德福德郡
伦敦卢顿机场

以利－里利研究中心
1992
温德尔莎姆，萨里郡
里利研究中心有限公司

布莱德维尔小区
1992
伦敦中央东4区
哈斯米尔房地产有限公司

*香港大屿山－机场铁路中央车站
1992
香港
公共交通铁道公司

本州造纸厂
1992
福冈，日本
东京 M·I·O 公司

*幼儿园，罗森海姆，法兰克福
1992
罗森海姆，法兰克福，德国
设计竞赛参赛作品

*杜塞尔多夫塔楼
1992
杜塞尔多夫，德国
设计竞赛参赛作品

*柏林 2000 年奥运会场馆
1992
柏林，德国
设计竞赛参赛作品

*废料利用发电厂
1992
贝尔威迪，肯特郡
科利环境有限公司

煤气公司总部
1992
伯明翰商业园区，英格兰中西部地区
英国煤气公司

劳埃德商船协会总部
1992
利普霍克，汉普郡
设计竞赛参赛作品

帝国战争博物馆二期工程
1993
伦敦东南1区
环境部

政府办公楼
1993
圣乔治大道，伦敦西南1区
公共资产机构

*里昂阿尔卑斯科技园区
1993
里昂，法国
里昂市政署

*伊斯坦布尔文化和会议中心
1993
伊斯坦布尔，土耳其
伊斯坦布尔文化艺术基金会

国家保龄球中心
1993
沃信，西苏塞克斯郡
沃信市自治议会

*格罗大楼，新城
1993
新城，威尔士
威尔士农村发展委员会/管理科技公司

*埃利斯公园田径运动场
1993
约翰内斯堡，南非
约翰内斯堡市政府

FPW 市场
1993
伊斯坦布尔，土耳其
FPW 供销合作社

*曼彻斯特 2000 奥林匹克国际体育场
1993
曼彻斯特
AMEC 投资开发公司

内部庭院开发工程
1994
英国博物馆，伦敦中央西1区
设计竞赛参赛作品

Awards & Exhibitions 获奖及参展作品

* Commendation

* 表示推荐

获奖作品

混凝土学会奖
英国混凝土学会
劳埃德银行大楼
1993

英国皇家建筑师学会区域规划奖和低地设计奖
英国皇家建筑师学会
法人－通用保险公司大楼
1992

商业环境贡献奖
基础基金会
* 斯托克利园区
1992

城镇发展贡献奖（Civic Trust Award）
城镇发展联合会（The Civic Trust）
皇家人寿保险大楼
1992

规划成就年度奖
皇家城镇规划学会
布鲁德门
纪念银质奖杯
1992

钢结构设计奖
英国钢结构建筑协会
白金汉宫大道 123 号
1992

英国皇家建筑师学会主席推荐"年度建筑"
英国皇家建筑师学会
布鲁德门
1991

建筑运行奖
财经时报
* 哈斯伯罗大楼
1991

英国皇家建筑师学会奖
英国皇家建筑师学会
苏塞克斯正面看台，古德伍德赛马场
1991

环境保护奖
皇家特许测量师协会
斯托克利园区
1991

城镇发展贡献奖
城镇发展联合会
* 科坡斯恩酒店
1991

城镇发展贡献奖
城镇发展联合会
帝国战争博物馆
1991

朗利梅德年度设计奖
朗利梅德区自治会
温特沃斯网球馆（1区）
1991

皇家特许测量师协会城市改造奖
皇家特许测量师协会
* 布鲁德门
1991

天然石材设计奖
英国采石业联合会
布鲁德门广场
1991

工艺与制造奖
沃平工艺联合会
斯托克利园区
1991

年度建筑奖
皇家美术委员会／星期日时报
* 苏塞克斯正面看台，古德伍德赛马场
1991

特别设计奖
西苏塞克斯乡村自治会
苏塞克斯正面看台，古德伍德赛马场
1990－1991

工艺与制造奖
沃平工艺联合会
奥雅纳事务所办公楼
1990

电气节能二等奖
电力学会
法人-通用保险公司大楼
1990

年度建筑奖
皇家美术委员会/星期日时报
帝国战争博物馆
1990

建筑设计施工创新奖
PA咨询集团
布鲁德门
1989

景观保护奖
伦敦旅行家协会
斯托克利园区
1989

斯蒂特利奖
城镇发展联合会
布鲁德门
1989

城镇发展贡献奖
城镇发展联合会
斯托克利园区
1989

工艺与制造奖
沃平工艺联合会
哈斯伯罗大楼
1989

建筑运行奖
财经时报
斯托克利园区
1989

商业区景观规划奖——包揽所有奖项
地产学报/杰克逊-斯托普斯机构
斯托克利园区
1989

钢结构设计奖
英国钢结构建筑协会
帝国战争博物馆
1989

欧洲钢结构建设联合会奖
布鲁德门
1988

钢结构设计奖
英国钢结构建筑协会
布鲁德门
1989

独立/年轻土木工程师工业建筑奖
英国建设工业行会
布鲁德门
1988

英国皇家建筑师学会奖
英国皇家建筑师学会
巴伯夫区自治会办公楼
1987

英国皇家建筑师学会奖
英国皇家建筑师学会
芬斯伯里大道1号
1987

英国皇家建筑师学会奖
英国皇家建筑师学会
*国际园艺节展厅
1987

Awards & Exhibitions Continued

建筑设计施工创新奖
PA 咨询集团
斯托克利园区
1987

规划成就年度奖
皇家城镇规划学会
斯托克利园区
1987

城市改造奖
欧洲环境年
*斯托克利园区
1987

城镇发展贡献奖
城镇发展联合会
布莱克里夫大楼
1986

罗伯特·马太奖
由英联邦建筑师学会二年一度颁发的奖项，授予对建筑有创新贡献的企业
1985

建筑运行奖
财经时报
芬斯伯里大道 1 号
1985

年度办公室设计奖
英国办公室委员会
布莱克里夫大楼
1985

钢结构设计奖
英国钢结构建筑协会
芬斯伯里大道 1 号
1985

城镇发展贡献奖
城镇发展联合会
国际园艺节展厅
1985

城镇发展贡献奖
城镇发展联合会
IBM 公司北港工程
1984

城镇发展贡献奖
城镇发展联合会
*巴伯夫区自治会办公楼
1984

钢结构设计奖
英国钢结构建筑协会
国际园艺节展厅
1984

英国皇家建筑师学会奖
英国皇家建筑师学会
特房伯有限公司工厂
1983

建筑运行奖
财经时报
盖特威 2 号楼
1983

钢结构设计奖
英国钢结构建筑协会
*IBM 公司北港工程
1983

城镇发展贡献奖
城镇发展联合会
*贝德福德学校
1982

商业和工业奖
英国环境工业专门小组
特房伯有限公司工厂
1982

英国皇家建筑师学会奖
英国皇家建筑师学会
托马斯·怀特爵士大楼，圣约翰学院，牛津
1981

英国皇家建筑师学会奖
英国皇家建筑师学会
伦敦劳埃德保险公司总部
1981

英国皇家建筑师学会奖
英国皇家建筑师学会
杜鲁门有限公司总部
1981

英国皇家建筑师学会奖
英国皇家建筑师学会
英国中央电力局西南区总部
1980

城镇发展贡献奖
城镇发展联合会
伦敦劳埃德保险公司总部
1980

城镇发展贡献奖
城镇发展联合会
英国中央电力局西南区总部
1980

建筑运行奖
财经时报
*英国中央电力局西南区总部
1980

商业和工业奖
英国环境工业专门小组
*伦敦劳埃德保险公司总部
1980

英国皇家建筑师学会奖
英国皇家建筑师学会
盖特威1号楼
1979

建筑运行奖
财经时报
皇家海军朴次茅斯基地码头大楼
皇家海事附属维护检修和供给中心
1979

商业和工业奖
英国环境工业专门小组
英国中央电力局西南区总部
1979

城镇发展贡献奖
城镇发展联合会
*盖特威1号楼
1978

建筑运行奖
财经时报
*IBM公司北港工程
1978

商业和工业奖
盖特威1号楼
1978

建筑运行奖
财经时报
杜鲁门有限公司总部
1977

钢结构设计奖
英国钢结构建筑协会
特拉法尔加住宅开发有限公司办公楼
1977

年度办公室设计奖
英国办公室委员会
杜鲁门有限公司总部
1977

混凝土学会奖
英国混凝土学会
托马斯·怀特爵士大楼,圣约翰学院,牛津
1976

商业和工业奖
英国环境工业专门小组
杜鲁门有限公司总部
1976

Awards & Exhibitions Continued

混凝土学会奖
英国混凝土学会
*安普福思学院教室及宿舍建筑
1975

英国皇家建筑师学会奖
英国皇家建筑师学会
*英国糖业公司中心办公室
1975

建筑运行奖
财经时报
皇家海军朴次茅斯基地厚钢板制造厂
1975

英国皇家建筑师学会奖
英国皇家建筑师学会
*剑桥大学动物学系、冶金系、计算机技术系
1974

英国皇家建筑师学会奖
英国皇家建筑师学会
*音乐学院，东英吉利亚
1974

英国皇家建筑师学会奖
英国皇家建筑师学会
*企鹅图书有限公司办公楼
1973

英国皇家建筑师学会奖
英国皇家建筑师学会
*牛津邮政时报印刷厂及办公楼
1973

城镇发展贡献奖
城镇发展联合会
*私人住宅，温布尔登
1973

建筑运行奖
财经时报
地平线项目，约翰·普莱耶和后裔的工厂
1973

城镇发展贡献奖
城镇发展联合会
*伯明翰大学文学院和商学院
1972

城镇发展贡献奖
城镇发展联合会
*勒斯特大学人文和社会科学院
1972

城镇发展贡献奖
城镇发展联合会
牛津邮政时报印刷厂及办公楼
1972

城镇发展贡献奖
城镇发展联合会
地平线项目，约翰·普莱耶和后裔的工厂
1972

建筑运行奖
财经时报
IBM（英国）计算机制造装配中心
1972

钢结构设计奖
英国钢结构建筑协会
IBM 公司北港工程
1970

混凝土学会奖
英国混凝土学会
萨默维尔学院,牛津
1969

混凝土学会奖
英国混凝土学会
*企鹅图书有限公司办公楼和书店

城镇发展贡献奖
城镇发展联合会
莫尔丁斯音乐厅
1968

城镇发展贡献奖
城镇发展联合会
奥夫·阿鲁普和合伙人事务所办公楼
1968

城镇发展贡献奖
城镇发展联合会
*沃恩大楼,萨默维尔学院,牛津大学
1966

英国皇家建筑师学会奖
英国皇家建筑师学会
伯明翰大学采矿和冶金学院
1966

城镇发展贡献奖
城镇发展联合会
圣体学院,雷克汉普顿楼,剑桥
1965

参展作品

皇家艺术院夏季展览
埃利斯公园田径运动场;废料利用发电厂;伊斯坦布尔文化和会议中心
皇家艺术院,伦敦
1994

皇家艺术院建筑展
曼彻斯特 2000 奥林匹克国际体育场;法人-通用保险公司大楼;苏塞克斯正面看台,古德伍德赛马场
皇家艺术院,纽约
1994

主席推荐作品展
伊斯坦布尔文化和会议中心
西英格兰皇家艺术院
布里斯托尔
1994

城市的变革:伦敦城市建筑展
*芬斯伯里大道 1 号
建筑基金会
伦敦,巴塞罗那,布拉格
1993

城市设计进展
联合大农场;布鲁德门;斯托克利园区
英国皇家建筑师协会
伦敦
1993

节能概念设计展
英国中央电力局西南区总部;盖特威 2 号楼;英国中央电力局西南区总部
国际建筑联合会,芝加哥
1993

体育馆设计的新方向
苏塞克斯正面看台,古德伍德赛马场;曼彻斯特 2000 奥林匹克国际体育场
建筑中心,伦敦
1993

区域设计观念展
皇家人寿保险资产有限公司;福布斯·梅隆图书馆,剑桥克莱尔学院;联合大农场
英国皇家建筑师学会
剑桥
1993

Awards & Exhibitions Continued

英国设计映像展
帝国战争博物馆
设计师特许协会
格拉斯哥
1993

国际建筑和钢结构研讨会
白金汉宫大道123号；苏塞克斯正面看台，古德伍德赛马场
建筑和钢结构研讨会
巴黎
1993

欧洲节能概念建筑展
皇家人寿保险资产有限公司；盖特威2号楼
都柏林大学学院
1993

节能概念设计展
幼儿园，罗森海姆，法兰克福
AEDES建筑艺术展
柏林
1992

建筑模型展
柏林2000年奥运会场馆
英国皇家建筑师学会
伦敦
1992

将英国建设得更美好
近门旅馆；帝国战争博物馆；法人－通用保险公司大楼
城镇发展联合会

建筑艺术
布鲁德门；帝国战争博物馆；斯托克利园区
布若克斯波恩仲夏节
布若克斯波恩
1992

国际建筑工业展
斯托克利园区
博洛尼亚博览会
博洛尼亚
1992/93

建筑中的工程师
建筑中心联合会
伦敦
1991

建筑秩序
法人－通用保险公司大楼；皇家人寿保险资产有限公司；劳埃德银行办公楼；苏塞克斯正面看台，古德伍德赛马场
布兰克·德·巴格斯陈列馆
伦敦
1991

帕特诺斯特工程成就展览
主祷广场
圣保罗大教堂
伦敦
1988

巴黎双年展
IBM科莎姆项目
巴黎市政府
巴黎
1985

CLAWSA房屋展
国际园艺节展厅；芬斯伯里大道1号；布莱克里夫大楼
CLAWSA
伦敦
1984

巡展
托马斯·怀特爵士大楼，牛津圣约翰学院；英国中央电力局西南区总部；IBM科莎姆项目
英国议会
伦敦，巴格达
1984

英国铁路展览会
英国中央电力局西南区总部；芬斯伯里大道1号；盖特威2号楼；布莱克里夫大楼
英国铁路局
沃特福德

现代建筑展
英国中央电力局西南区总部；埃弗的私人住宅；巴伯夫区自治会办公楼
沃里克艺术联合会
伦敦
1984

环境与建筑
英国中央电力局西南区总部
科技博物馆，环境设施陈列室
伦敦
1984

建筑艺术
IBM科莎姆项目
英国皇家建筑师学会
伦敦
1983

英国建筑展
国际园艺节展厅
英国皇家建筑师学会
伦敦
1983

获奖作品展
IBM科莎姆项目；特庑伯有限公司工厂；芬斯伯里大道1号
英国皇家建筑师学会
伦敦
1983

威尼斯双年展
市中心巴伯舍夫住宅区及商业区
威尼斯双年展
威尼斯
1982

1982年英国建筑展
国际园艺节展厅
英国皇家建筑师学会/《建筑设计》期刊
伦敦
1982

英国皇家建筑师协会获奖作品展
托马斯·怀特爵士大楼，牛津圣约翰学院；伦敦劳埃德保险公司总部
英国皇家建筑师学会
伦敦
1981

建筑与设计
莫尔丁斯音乐厅；盖特威1号楼；伦敦劳埃德保险公司总部；英国中央电力局西南区总部；特庑伯有限公司工厂
巴兹尔登开发公司
巴兹尔登
1981

肯特郡自治会获奖作品展
伦敦劳埃德保险公司总部
城镇发展联合会
肯特
1981

英国建筑选展
贝德福德学校；IBM科莎姆项目；特庑伯有限公司工厂
英国皇家建筑师学会
伦敦
1981

工业建筑研讨会
IBM（英国）计算机制造装配中心；朴次茅斯码头
英国皇家建筑师学会/全国建筑企业贸易联合会（NFBTE）
伦敦

室内设计展
杜鲁门有限公司总部；盖特威1号楼
建筑中心
伦敦
1980

Awards & Exhibitions Continued

帕尔马展览
托马斯·怀特爵士大楼，牛津圣约翰学院
帕尔马市政署
帕尔马
1979

常设性展览
托马斯·怀特爵士大楼，牛津圣约翰学院；
杜鲁门有限公司总部；盖特威1号楼；英国
中央电力局西南区总部
英国皇家建筑师学会客户咨询机构
伦敦
1979

英国皇家建筑师学会获奖作品展
盖特威1号楼；英国中央电力局西南区总部
英国皇家建筑师学会
中东，意大利，匈牙利，南美
1980

布克斯顿节
布克斯顿歌剧院
布克斯顿节
布克斯顿
1979

英国皇家建筑师学会节能设计研讨会
英国中央电力局西南区总部
英国皇家建筑师学会
伦敦
1979/80

新建筑
顾问建筑师学会
皇家艺术院沙龙，伦敦
1978，1979，1982，1984

英国建筑
托马斯·怀特爵士大楼，牛津圣约翰学院
英国议会
德黑兰
1978

英国皇家建筑师学会
托马斯·怀特爵士大楼，牛津圣约翰学院；
安普福思学院；杜鲁门有限公司总部；伦敦
劳埃德保险公司总部
纽卡斯尔
1978

建筑联合学院校庆纪念展
托马斯·怀特爵士大楼，牛津圣约翰学院；
安普福思学院；杜鲁门有限公司总部；地平
线项目，约翰·普莱耶和后裔的工厂
建筑联合学院
伦敦
1977

模型展
托马斯·怀特爵士大楼，牛津圣约翰学院；
布什·莱恩大楼
谢菲尔德建筑师学会
谢菲尔德
1974

阿鲁普联合事务所展览
建筑联合学院
伦敦
1973

电气工程师协会展览
地平线项目，约翰·普莱耶和后裔的工厂
电气工程师协会
伯爵府邸（Earls Court），伦敦
1972

Bibliography

参考文献

General Publications

"Arup Associates, Architects and Engineers in Inghilterra, due Edific Industriali." *Domus* (Milan, July 1973): pp. 4–10.

"Big Bang, Big Boom." *Business* (July 1987): pp 60–75.

Blanc, A., McEvoy, M. & Plank, R. (eds). *Architecture and Construction in Steel*. London: Steel Construction Institute/E. & F.N. Spon, 1993, pp. 1–10, 121–130, 289–320.

Brawne, M. *Arup Associates. The Biography of an Architectural Practice*. London: Lund Humphries, 1983.

Brookes, A. J. *Concepts in Cladding. Case Studies of Jointing for Architects and Engineers*. London: Construction Press, 1985, pp. 16–19, 32–36.

Carter, B. "Coming to Terms with Climate." *Architectural Review* (June 1991): pp. 84–87.

Carter, B. "Mastering the Multi-Disciplinary." *World Architecture* (no. 11, 1991): pp. 72–76.

Carter, B. "The Office Outdoors." *Landscape Design* (May 1992): pp. 14–16.

Carter, B. & Warburton, P. "The Development of a Solar Architecture." *Detail* (Munich, vol. 6, 1993): pp. 671–675.

Davey, P. "Architecture Versus Big Business." *Architectural Review* (August 1988): pp. 15–19.

Dowson, P. "Arup Associates." Architecture Intérieure (September/October 1979): pp. 80–88.

Dowson, P. "Arup's Office Evolution." *Architectural Review* (May 1992): pp. 54–55.

Forsyth, M. *Auditoria: Designing for the Performing Arts*. London: The Mitchell Publishing Company Limited, 1987.

Frewer, R. "Planning and Designing the Place of Work." In Wildenmann, R. (ed.). *Stadt, Kultur, Natur: Chancen zukunftiger Lebensgestaltung. (City, Civilisation, Nature: Opportunities for Planning Life in the Future.)* Baden-Baden: Nomos, 1989, pp. 391–398.

Glancey, J. "Building's Big Bang." *Management Today* (February 1988): pp. 85–98.

Glancey, J. *New British Architecture*. London: Thames and Hudson, 1989.

Hanson, M. "Architects Today. Arup Associates." *Estates Gazette*. (July 1989): pp. 20–21, 60.

SD Editorial Division *High-tech*. Tokyo: Kajuna Institute of Publishing Co. Ltd, 1987, pp. 74–77.

Hindhaugh, E.W. (ed.). *Building with British Steel*. English edition. British Steel, 1990, pp. 7, 46, 48, 80.

Lord, P. & Templeton, D. *Architecture of Sound: Designing Places of Assembly*. London: Architectural Press, 1986.

Lush, D.M. "L'edificio Intelligente. Intelligent Buildings." *l'Arca* (Milan, April 1987): pp. 48–57.

Lyall, S. *The State of British Architecture*. London: Architectural Press, 1980.

MacCormac, R. "The Dignity of Office." *Architectural Review* (May 1992): pp. 76–82.

McKean, C. & Jestico, T. *Editors' Guide to Modern Buildings in London 1965–75*. London: Warehouse Publishing Ltd, 1976.

Moffett, N. *The Best of British Architecture 1980–2000*. London: E. & F.N. Spon, 1993, pp. 21, 56.

Moxley, R. & Woods, F. (eds). *Illustrated Directory of Architects 1984*. London: Association of Consultant Architects/Architectural Press, 1984.

Murray, P. & Trombley, S. (eds). *ADT Modern Architecture Guide: Britain*. London: Architecture Design & Technology Press, 1990.

Murray, P. & Trombley, S. (eds). *Modern British Architecture Since 1945*. London: RIBA, 1984.

Ogg, A. *Architecture in Steel: The Australian Context*. Redhill, ACT: Royal Australian Institute of Architects, 1987.

Orton, A. *The Way We Build Now: Form, Scale and Technique*. Wokingham: Van Nostrand Reinhold, 1988: pp. 525–530.

Pawley, M. "Make Me an Office." *Tatler* (June 1987): pp. 126–127, 156, 158.

Phillips, A. *The Best in Industrial Architecture*. London: B.T. Batsford Ltd, 1993.

Jencks, C. "Post-Modern Construction and Ornament." *Architectural Design* (no. 11/12, 1990): pp. 36–43.

Powell, K. *World Cities: London*. London: Academy Editions, 1993.

Rawle, T. *Cambridge Architecture*. Cambridge: Trefoil Books, 1985, pp. 65, 66, 106–109, 193, 201.

Roda, R. "Arup Associates: Una Comunita'di Progettisti." *Modulo* (Florence, March 1990): pp. 198–209.

Saxon, R. *Atrium Buildings. Development and Design*. London: Architectural Press, 1986.

Sharp, D. *20th Century Architecture: A Visual History*. London: Lund Humphries, 1991.

Sugden, D. "The Skeletal Frame from Arkwright to Arup." *Baukultur* (Hamburg, no. 5, 1986): pp. 3–14.

Sugden, D. (contributor) "25 Years of British Architecture 1952–1977." *RIBA Journal* (May 1977): p. 190.

"Wooden Architecture Today." *Space Design* (Tokyo, January 1987).

"Working it Out." *Architectural Review* (May 1992): pp. 56–61.

Factory at Hemel Hempstead

"Factory at Hemel Hempstead." *Architects' Journal* (20 December 1956): pp. 901–914.

"Factory, Hemel Hempstead." *Architects' Journal* (vol. 138, no. 18, 1963): pp. 924–926.

"Factory at Hemel Hempstead." *Architectural Review* (April 1957): pp. 230–234.

"Factory, Hemel Hempstead. Working Details of Factory Roof." *Architects' Journal* (February 1957): pp. 340–341.

Fry, M. "Factory at Hemel Hempstead." *Zodiac* (vol. 1, 1957): pp. 181–184.

Bibliography Continued

"Working Detail Illustrating Factory Cladding." *Architects' Journal* (vol. 123, no. 32, 1956).

Research Laboratories and Maintenance Building for CIBA, Duxford

"Estension d'usine à Duxford." *l'Architecture Française* (Paris, August 1960): pp. 64–67.

"Factory Extensions at Duxford." *Architectural Review* (November 1959): pp. 254–260.

"Laboratories and Offices." *Architects' Journal* (vol. 140, no. 24, 1964): pp. 1175–1192, 1194.

"Laboratories and Process Block." *Architects' Journal* (vol. 131, no. 3378, 1960): pp. 79–88.

McKean, C. *Architectural Guide to Cambridge and East Anglia Since 1920.* Cambridge: RIBA, 1982.

"Nouveau Bâtiment des Laboratoires CIBA à Duxford, Cambridge." *l'Architecture d'Aujourdhui* (April/May 1961): pp. 44–45.

Vaughan Building, Somerville College, Oxford

"Architetture Universitaire Inglesi – due Progetti per Oxford." *Domus* (Milan, no. 401, 1963): pp. 2–7.

"Graduate Building, Somerville College, Oxford." *Architectural Review* (vol. 137, no. 816, 1965): pp. 110–113.

"Somerville Builds for Graduates." *Architects' Journal* (vol. 140, no. 18, 1964): p. 972.

"Somerville College, Oxford." *Architects' Journal* (vol. 130, no. 33, 1959) pp. 495–496.

"Somerville College, Oxford." *Werk, Bauen und Wohnen* (Zurich, no. 8, 1962): pp. 359–60.

Point Royal Flats, Bracknell, Berkshire

"Bracknell's Campanile." *Architects' Journal* (vol. 138, no. 16, 1963).

"Imeuble – Tour a Bracknell." *l'Architecture d'Aujourdhui* (no. 120, 1965): pp. 36–37.

"Point Royal, Bracknell." *l'Architecture d'Aujourdhui* (April/May 1964): p. 135.

"Point Royal Flats." *Architects' Journal* (vol. 139, no. 20, 1964): pp. 1099–1112.

"Point Royal Flats." *Architectural Design* (vol. 33, no. 11, 1963): p. 503.

"Point Royal Flats." *Architecture & Urbanism* (Tokyo, December 1977): p. 85.

"Point Royal Flats, Bracknell, Corpus Christi College, Birmingham University." *Kenchiku Bunka* (Tokyo, 1966): pp. 79–90.

"Urban Design in Practice." *Urban Design Quarterly* (September 1991): pp. 12–13.

Factory for York Shipley Ltd, Basildon, Essex

"Deux Systèmes de Couverture Translucide." *l'Architecture d'Aujourdhui* (Paris: September 1967): p. xxi.

"Factory at Basildon New Town." *Architects' Journal* (vol. 137, no. 1, 1963): pp. 33–42.

"Una Fabbrica a Basildon." *Ediliza Moderna* (no. 78, 1963): pp. 23–26.

"York Shipley, Basildon." *Industrial Architecture* (April 1963): pp. 238–239.

Nuclear Physics Laboratory, Oxford University

Bendixson, T.M.P. "Oxford's New Physics Laboratories." *Architects' Journal* (vol. 133, no. 34, 1961): pp. 837–838.

Calder, N. "A New Machine to Explore the Atomic Nucleus." *New Scientist* (vol. 10, no. 238, 1961): pp. 571–574.

"New Physics Building, Oxford University." *Werk, Bauen und Wohnen* (Zurich, no. 8, 1962): pp. 355–358.

Swimming Pool, Walton-on-Thames

"Hallenbad Walton-on-Thames." *Glasform* (September/October 1967): pp. 22–25.

Metcalf, P. "Four British Swimming Baths." *Architectural Review* (vol. 141, no. 843, 1967): pp. 345–346.

"Swimming Baths at Walton-on-Thames." *Architects' Journal* (vol. 143, no. 2, 1966): pp. 91–104.

"Walton-on-Thames Swimming Pool." *Deutsche Bauzeitung* (Stuttgart, vol. 101, no. 1, 1967): pp. 18–21.

Laboratories for CIBA, Duxford

"Factories at Duxford, Cambs." *Architects' Journal* (vol. 143, no. 14, 1967): pp. 902–907.

"Laboratoires CIBA, Duxford, Grande Bretagne." *l'Architecture d'Aujourdhui* (September 1967): p. xxiii.

Mining & Metallurgy Department, Birmingham University

Alexander, A. "A Building as a System." *Architectural Forum* (vol. 125, no. 1, 1966): pp. 90–97.

Carter, B. "Structural Rationalism." *Architects' Journal* (vol. 175, no. 3, 1982): p. 68.

Dowson, P. "Building for Science: University of Birmingham." *Architectural Design* (vol. 37, no. 4, 1967): pp. 160–170.

"Laboratories, Mining and Metallurgy, University of Birmingham." *Architects' Journal* (vol. 145, no. 15, 1967): pp. 905–918.

"Panorama University Buildings." *l'Architecture d'Aujourdhui* (April/May 1968): p. 38.

"RIBA Award West Midlands Region." *Architects' Journal* (vol. 144, no. 1, 1966): p. 2.

"The Secret Gardens of Birmingham." *Architects' Journal* (vol. 152, no. 28, 1970): pp. 110–158.

"University of Birmingham." *Architectural Design* (vol. 36, no. 10, 1966): p. 479.

"University of Birmingham." *Zodiac* (no. 18, 1968): pp. 72–77.

"University of Birmingham, Department of Mining & Metallurgy." *Architecture & Urbanism* (Tokyo, December 1977): pp. 99–103.

Yamashita, K. "Birmingham University: Mining, Metallurgy & Mineral Engineering Department." *Japan Architect* (vol. 45, no. 7, 1970): pp. 92–94.

Leckhampton House, Corpus Christi College, Cambridge

Brandenburger, J. "The Craven Image, or The Apotheosis of the Architectural Photograph." *Architects' Journal* (vol. 170, no. 131, 1979): p. 239.

"Building Revisit: Graduate Housing at Cambridge." *Architects' Journal* (vol. 157, no. 11, 1973): pp. 607–618.

"Cambridge Contrasts." *Architects' Journal* (vol. 147, no. 26, 1968): pp. 1426–1451.

"Corpus Christi College, Cambridge." *Architectural Design* (36: 12, 1966): p. 596.

Dowson, P. "A Room of One's Own." *Architectural Design* (vol. 38, no. 4, 1968): pp. 164–172.

"Graduate Building, Corpus Christi College, Cambridge." *Architectural Review* (vol. 137, no. 816, 1965): pp. 105–110.

"Leckhampton House des Corpus Christi College, Cambridge." *Werk, Bauen und Wohnen* (Zurich, vol. 53, no. 1, 1966): pp. 23–25.

"Residential Building for Corpus Christi College, Cambridge." *Architects' Journal* (vol. 141, no. 7, 1965): pp. 411–424.

"Studentenheim des Corpus Christi College, Cambridge." *Architektur und Wohnform* (Stuttgart, vol. 75, no. 7, 1967): pp. 466–467.

Factory for Walter Jones & Co., Ashford, Kent

"Factories at Ashford, Kent." *Architects' Journal* (vol. 143, no. 14, 1966) pp. 893–901.

"Factory for Electrical Components, Ashford." *International Asbestos Cement Review* (October 1969): p. 53.

Hobbs, R.W. "Services, Structure & Building." *Institution of Heating and Ventilation Engineers Journal* (February 1968): pp. 325–335.

Sugden, D. "The Anatomy of the Factory." *Architectural Design* (vol. 38, no. 11, 1968): pp. 513–551.

House, Drax Avenue, Wimbledon, London

"House at Wimbledon." *Architecture & Urbanism* (vol. 3, no. 12, 1974): pp. 34–37.

"House at Wimbledon." *Architectural Review* (vol. 152, no. 906, 1972): pp. 83–88.

"House for J. Zunz." *Architecture & Urbanism* (December 1977): pp. 140–141.

Attenborough Building, Leicester University

Di Luzio, C. "L'Attenborough Building per la Facolta di Arti e Science Sociali dell'Universita di Leicester." *Industria Italiano del Cemento* (October 1977): pp. 841–852.

Offices for Oxford Mail & Times

Ackermann, K. *Building for Industry.* Watermark Publications (UK) Limited, 1991.

Girouard, M. "Newspaper Offices, Oxford." *Architectural Review* (April 1972): pp. 223–232.

New Museums, Cambridge

Crosby, T. "New Museums Building, Cambridge." *Architectural Review* (February 1974): pp. 71–84.

The Maltings Concert Hall

Forsyth, M. *Buildings for Music.* Cambridge: Cambridge University Press, 1985.

Sugden, D. "Back to the 'Shoe Box'." *Architects' Journal* (March 1992): pp. 20–23.

Boulton House, Trinity Hall, Cambridge

"Boulton House, Trinity Hall, Cambridge." *Architectural Review* (August 1969): pp. 98–101.

"Boulton House, Trinity Hall, Cambridge." *Concrete* (July–September 1969): pp. 10–15.

Horizon Factory for John Player & Son

Hawes, F. "Player's Horizon." *Concrete Quarterly* (Winter 1988): pp. 6–9.

"Player's Horizon Project." *Financial Times* (2 November 1971): pp. 37–39.

IBM (UK) Ltd Head Office

"IBM United Kingdom Ltd." *Architecture & Urbanism* (3: 5, 1973) pp. 5–16.

Wright, L. "IBM Havant: Factory and Offices." *Architectural Review* (January 1972): pp. 5–14.

Sir Thomas White Building, St John's College, Oxford

Di Luzio, C. "Il Sir Thomas White Buildings per il St John's College ad Oxford, Gran Bretagna." *L'Industria Italiana del Cemento* (Rome, vol. 49, no. 9, 1979): pp. 519–532.

"Oxford: New Buildings at Keble and St John's." *Architectural Review* (December 1977): pp. 349–364.

Bibliography Continued

"Sir Thomas White Building, Oxford." *Architecture & Urbanism* (Tokyo, December 1977): pp. 77–84.

"Students' Homes in Oxford." *Baumeister* (Munich, vol. 75, no. 5, 1978): pp. 415–418.

Bush Lane House, London

Murphy, S.J. *Continuity and Change: Building in the City of London*. London: Corporation of London, 1984.

Headquarters, Truman Ltd, London

"A Brewery between Georgian Houses." *Bauen und Wohnen* (no. 3, 1979): pp. 74–76.

"Brick Lane Brewery, E1." *RIBA London Region Yearbook*. London: RIBA, 1981, pp. 88–89.

Cantacuzino, S. "Brick Lane Brewery." *Architectural Review* (April 1978).

"Truman's Brewery, 91 Brick Lane, London E1." *Concrete Quarterly* (April/June 1980): pp. 8–11.

Scottish Opera, Glasgow

"Arup: Opera House." *Architectural Review* (April 1976): pp. 202–207.

Macpherson, H. "Scottish Opera Finds a New Home." *Observer Magazine* (12 October 1975): pp. 47–49.

"Operatic Miracle." *Architects' Journal* (22 October 1975): pp. 833–834.

"Scotland's First Opera House Opens Tonight." *The Scotsman* (14 October 1975): pp. 12–13.

Stevens Curl, J. "Scottish Opera at Home." *Country Life* (6 November 1975): p. 1213.

Sugden, D. "Scottish Opera at the Theatre Royal Glasgow." *TABS Stage Lighting International* (Summer 1976): pp. 3–7.

Gateway 1, Basingstoke

"Building Study: Gateway House, Basingstoke." *Architects' Journal* (24 August 1977): pp. 343–358.

"Inside the Office: A Look at the Changes Taking Place in Office Building." *Architects' Journal* (19 & 26 August 1987): pp. 33–81.

Oswalt, P. & Rexroth, S. *Wohltemperierte Architektur.* (Well-tempered Architecture.) Heidelberg: Verlag C.F. Müller GmbH, 1994, pp. 184–189.

Scrivens, S. "Landscape Revisit 2: Landscape Management Gateway 1 House." *Architects' Journal* (11 December 1985): pp. 41–47.

"Verwaltungsbau in Basingstoke." *Baumeister* (Munich, October 1978): pp. 862–865.

Administrative Headquarters, Lloyd's of London, Chatham

"Administrative Headquarters, Chatham." *Architecture & Urbanism* (Tokyo, September 1981): pp. 48–53.

"Arup's Constructed Void." *Architects' Journal* (4 February 1981): pp. 198–217.

Dinelli, F. "Lloyd's Chatham Administrative Headquarters." *L'Industria Italiana del Cemento* (Rome, April 1981): pp. 42–45.

"Lloyd's Administrative Headquarters, Chatham." *Detail* (Nuremberg, March/April 1984): pp. 169–170.

"Lloyd's Administrative Headquarters in Chatham." *Architectural Design* (vol. 51, issue 3/4, 1981): pp. 7–9.

"Lloyd's in Chatham." *Baumeister* (Munich, vol. 77, no. 9, 1980): pp. 886–889.

"Sur le Porte de Chatham." *Architecture Interieure* (Paris, Sept/Oct 1980): pp. 120–122.

Von Gerkan, M. (ed.). *Geneigte Dacher: Tragwerke.* (Pitched Roofs: Structure.) Berlin: Edition Detail/Rudolf Muller, 1989.

CEGB South West Region Headquarters, Bristol

"CEGB South West Regional Headquarters, Bristol." *Architecture & Urbanism* (Tokyo, September 1981): pp. 54–58.

Duffy, F. "Interior Motives." *Architects' Journal* (19 September 1979): pp. 591–605.

Hawkes, D. & Duffy, F. "Regional Headquarters for the CEGB South Western Region, Bristol." *Architects' Journal* (15 August 1979): pp. 325–343.

Hawkes, D. & Owers, J. *The Architecture of Energy*. Martin Centre for Architectural and Urban Studies, Cambridge University. Harlow: Construction Press, 1981, pp. 149–156.

MacCormac, R. "Offices for CEGB, Bristol." *Architectural Review* (July 1979): pp. 9–22.

Factory for Trebor Ltd, Colchester

"Agents of Industry: Trebor, Colchester." *Progressive Architecture* (Stamford, CT, December 1991): pp. 64–72.

Briarcliff House, Farnborough

"The Arup Approach: Underfloor Air Conditioning." *Building Services* (6: 8, 1984): pp. 33–34.

Ashley, S. "Air up from Under: Underfloor Air Conditioning Systems." *Building Services* (October 1988): pp. 29–31.

Commission of the European Communities "Briarcliff House." *Project Monitor* (Issue 12, December 1987).

"Evaluation of the Passive Solar Energy Design of Briarcliff House, Farnborough." *Journal of the Chartered Institute of Building Services* (January 1989): pp. 63–64.

Hannay, P. "Farnborough Salute." *Architects' Journal* (5 September 1984): pp. 63–69.

"Intelligent Buildings." *Building Design Supplement* (May 1988): pp. 12–28.

Nelson, G. "The Glass of '84. Briarcliff House." *Architects' Journal* (5 September 1984): pp. 70–82.

Nelson, G. "Energy Revisit. Briarcliff House." *Architects' Journal* (12 February 1986): pp. 53–56.

Buxton Opera House

"Buxton Opera House." *Architecture & Urbanism* (Tokyo, October 1981): pp. 88–91.

LeCuyer, A.W. "Listening Space: A Conversation with Derek Sugden of Arup Associates." *Building Design* (25 January 1980): pp. 20–21.

"Opera Revival – Buxton Opera House." *Architects' Journal* (169: 120, 1979): pp. 569, 579.

Sugden, D. "The Maltings Concert Hall, Buxton Opera House and York Minster. A Reconstruction and Two Restorations." *In Construction: A Challenge for Steel* (Conference proceedings, Luxembourg, September 1981): pp. 223–244.

Digital Equipment Co., Reading

Allsopp, K. "Digital Grid: Building Study, DEC Reading." *Architects' Journal* (vol. 177, no. 18, 1983): pp. 51–66.

Gateway 2, Basingstoke

"Architecture at Work: 1983 Award for Industrial and Commercial Building." *Financial Times* (6 December 1983): p. 28.

Clementson, N.R. "Computers Controlled." *Journal of the Chartered Institution of Building Services* (vol. 5, no. 4, 1983): pp. 38–39.

Cruickshank, D. "Demanding Design." *Architects' Journal* (vol. 176, no. 27, 1982): pp. 44–49.

"Financial Gains: Discussion of Financial Times Architecture at Work Award Winner." *Architects' Journal* (vol. 178, no. 49, 1983) pp. 32–35.

Hannay, P. "Building Study: Rooms with a View: Gateway 2, Basingstoke." *Architects' Journal* (14 November 1984): pp. 55–66.

Hawkes, D. "Air Apparent: Gateway 2." *Architects' Journal* (3 August 1983): pp. 26–34.

Hawkes, D. "Energy Revisit: Gateway 2." *Architects' Journal* (14 November 1984): pp. 72–73.

International Energy Agency/Databuild. *Passive and Hybrid Solar Commercial Buildings*. Harwell: Renewable Energy Promotion Group, 1991, pp. 227–232.

"1975–1985: The Years of Uncertainty: Gateway 2." *Concrete Quarterly* (January–March 1985): pp. 38–45.

Seager, A. "Energy Performance Assessments: Gateway 2, Basingstoke." *EPA Non-domestic Technical Report: Solar Building Study*. Birmingham: Databuild Ltd, 1992.

Shepherd, J. "Arup Associates' Scenic Lift at Basingstoke." *RIBA Journal* (March 1984): pp. 57–60.

Yannas, S. "Designing Buildings to Use Solar Energy." *Energy World* (September 1990): pp. 22–24.

Bab Al Sheikh, Baghdad

"Arup Associates: Urban Design, Baghdad, Iraq." *Architectural Review* (vol. 171, no. 1019, 1982): p. 58.

"Architettura nei Paesi Islamici. Seconda Mostra Internazionale di Architeturra." Venice: Edizioni la Biennale di Venezia, 1982, pp. 163–164.

"Bab Al Sheikh." *AD British Architecture* (1982): pp. 112–113.

"Bab Al Sheikh." *Architects' Journal* (vol. 73, no. 21, 1981): p. 987.

Mutschler, C. "Bab Al Sheikh." *Architekturexport und Architekturimport* (September 1982): pp. 388–391.

Diplomatic Quarter Sports Club, Riyadh

Amin, N. "A City to Cement Foreign Relations." *Saudi Arabia Review* (1983).

1 Finsbury Avenue, London

Amery, C. "An Objective View: A Survey of RIBA Award-winning Schemes." *RIBA Journal* (January 1988): pp. 26–28.

"Arup & Hopkins Win Financial Times Award." *Building Design* (6 December 1985): p. 5.

"British Architecture Now." *RIBA Journal* (vol. 90, no. 11, 1983): pp. viii–ix.

Brookes, A.J. & Grech, C. *The Building Envelope. Applications of New Technology Cladding*. London: Butterworth, 1990, pp. 41–45.

Burdett, R. (ed.). *City Changes: Architecture in the City of London 1985–1995*. London: Architecture Foundation/Corporation of London, 1992, pp. 25–28.

Carolin, P. "Racing up the Avenue." *Architects' Journal* (24–31 August 1983): pp. 65–67.

Davies, C. "Craft or Calculation?" *Architectural Review* (May 1985): pp. 19–30.

Davies, C. & Pearman, H. "Case Study: One Finsbury Avenue." *Designers' Journal* (January 1985): pp. 28–38.

"Finsbury Avenue." *New Civil Engineer* (21 November 1985): p. 9.

Foggo, P. "Fast Building." *Architects' Journal* (11 October 1989): pp. 77–81.

Holden, R. "Green Fringe in the City. Landscaping at the Broadgate Development and Finsbury Avenue." *Landscape Design* (April 1992): pp. 10–12.

Krafft, A. (ed.). *Contemporary Architecture: International Annual Review*. vol. 7. Lausanne: Bibliotheque des Arts, 1985.

"1 Finsbury Avenue in London." *Baumeister* (Munich, October 1985): pp. 43–49.

Pawley, M. "Declino e Caduta dell Architetture Londinese." ("The Decline and Fall of Architecture in London.") *Casabella* (Milan, September 1985): pp. 16–21.

"Solar Design Can Help Cut Britain's Heat and Light Bill." *Review* (April 1988): pp. 3–5.

Bibliography Continued

Spring, M. "Stretching City Limits." *Building* (14 October 1988): pp. 41–48.

"Steel and Fire Safety: A Global Approach." Eurofer, Steel Promotion Committee, 1990, p. 19.

Thornton, V. "Capital Architecture: An Introduction. London: An Architectural Guide." *RIBA Journal* (May 1991): p. 34.

Von Gerkan, M. (ed.). *Fassaden. (Facades.)* Cologne: Rudolf Muller, Edition Detail, 1988, pp. 89–92.

Forbes Mellon Library, Clare College, Cambridge

"Cambridge Conundrum." *Architects' Journal* (vol. 176, no. 42, 1982): pp. 32–33.

"Cambridge to Decide on Design for Clare College." *Architects' Journal* (vol. 176, no. 40, 1982) p. 55.

"Clare College, Cambridge: Forbes Mellon Library." *Concrete Quarterly* (July–September 1987): pp. 14–17.

"Cutting off a Lifeline." *Observer* (31 October 1982).

"Rival Verdicts on a Valued Varsity Vista." *The Guardian* (12 October 1982).

International Garden Festival Hall, Liverpool

"Arup Drome: Liverpool Garden Festival Hall." *Architectural Review* (July 1984): pp. 29–31.

"Beinahe ein Kristallpalast." ("Liverpool Garden Festival Hall.") *Deutsche Bauzeitung* (November 1984): pp. 19–23.

"British Architecture Now." *Building Design* (14 October 1983): pp. 26–43.

"British Architecture Now." *RIBA Journal* (vol. 90, no. 11, 1983): p. xiv.

Cass, R. "Liverpool 1984: The International Garden Festival." *Landscape Design* (October 1983): pp. 17–20.

"Centrepiece for International Garden Festival." *Civil Engineering Technology* (vol. 8, no. 10, 1984): p. 1.

"Crisis Time for Arup's Festival Hall." *Architects' Journal* (vol. 180, no. 32, 1983): p. 13.

"Edificio Polivalente a Liverpool." ("Liverpool Garden Festival Hall.") *l'Industria delle Construzione* (Milan, June 1985): pp. 52–56.

"Festival News." *Landscape Design* (August 1983): p. 7.

Frewer, R. "The Loose-fit Factor." *Architects' Journal* (176: 29, 1982): pp. 38–40, 52.

"Fringe Benefit – Liverpool's Festival Hall." *Architects' Journal* (vol. 180, no. 30, 1984): pp. 4–7.

"Garden Folly." *Architects' Journal* (vol. 181, no. 22, 1985): pp. 20–21.

"Garden of Eden on a Reclaimed Rubbish Dump." *RIBA Journal* (vol. 89, no. 9, 1982): p. 42.

Gossel, P. & Leuthauser, G. *Architektur des 20. Jahrhunderts. (Architecture of the 20th Century.)* Cologne: Benedikt Taschen, 1990, p. 337.

Hayward, D. "Festival Plots Slot in for Opening." *New Civil Engineer* (26 April 1984): pp. 28–31.

"High-tech." *Space Design* (Tokyo, January 1985): pp. 74–77.

"Industrial News: Structural Steel Design Awards 1984." *Structural Engineer* (vol. 63A, no. 1, 1985): p. 3.

"Liverpool '84. International Garden Festival Hall." *Baumeister* (Munich, July 1984): p. 12.

"Liverpool Garden Festival Hall." *Building Technology & Management* (vol. 22, no. 4, 1984): p. 18.

"Liverpool International Garden Festival 1984." *Landscape Design* (Special Issue, April 1984).

"Liverpool's Flower Festival." *Progressive Architecture* (Stamford, CT, vol. 66, no. 7, 1984): pp. 23–24.

Lyall, S. "Teddy Bears' Picnic." *New Society* (vol. 68, no. 1119, 1984): pp. 184–185.

Medhurst, J. "Liverpool Garden Festival Hall." *Architectural Review* (August 1984): pp. 71–80.

"Merseyside Garden Festival." *Economist* (19 February 1983).

"The Mersey Beat Brings Art Alive." *Sunday Times* (30 January 1983).

"Merseyside Initiative Confirmed." *Architects' Journal* (vol. 178, no. 28, 1983): p. 28.

"Metal-bending for Construction." *Architects' Journal* (vol. 181, no. 8, 1985): pp. 83–88.

Nooshin, H. (ed.). *International Conference on Space Structures No. 3, 11–14 September 1984*. Conference proceedings. Elsevier Applied Science, 1984, pp. 919–928.

Parker, D. & Bradshaw, T. "Another Look at Liverpool Garden Festival." *Landscape Design* (June 1986): pp. 32–35.

"Paxton's Flunkern: Festival Building in Liverpool." *Werk, Bauen und Wohnen* (Zurich, 1985): pp. 40–45.

Perkins, G. "Laurels for Liverpool." *Concrete Quarterly* (April–June 1984): pp. 14–26.

Raggett, T. "The Design and Construction of the Liverpool International Garden Festival Building." In Makowski, Z.S. (ed.). *Analysis, Design and Construction of Braced Barrel Vaults*. London: Elsevier, 1985, pp. 382–393.

Rushton, R. "Liverpool Festival Hall." *Royal Society of Arts Journal* (vol. 132, no. 537, 1984): pp. 633–634.

"Sowing £13m Seeds of Hope in Liverpool." *The Guardian* (12 July 1982).

Spring, M. "Urban Growth." *Building* (vol. 242, no. 23, 1982): p 27.

Stevens, T. "Glazed Arcades." *RIBA Journal* (March 1984): pp. 30–34.

Stiles, R. "Green Cities." *Architectural Review* (June 1984): pp. 23–31.

Teller, N. (ed.). *British Architectural Design Awards 1983*. London: Templegate Publishing, 1983.

"Urban Parks." *Landscape Design* (December 1992): pp. 9–47.

123 Buckingham Palace Road, Victoria, London

"123 Buckingham Palace Road." *New Steel Construction* (December 1992): p. 29.

Foggo, P. & Swenarton, M. "Walls for all Seasons." *Building Design* (3 June 1988): pp. 24–25.

Dawson, S. "Theme: Internal Elements. Case Study. Professional Class. 123 Buckingham Palace Road." *AJ Focus* (August 1991): pp. 15–19.

"Is This the New Victoriana?" *Building Design* (8 July 1983): p. 3.

Imperial War Museum, London

Atterbury, P. "Rival to the Tate." *Country Life* (27 April 1989): pp. 142–145.

"Back to the Front: Imperial War Museum." *The Times* (24 June 1989): p. 29.

Brawne, M. "War Games." *RIBA Journal* (September 1989): pp. 80–85.

Bunn, R. "Armed Services: Imperial War Museum." *Building Services* (August 1989): pp. 46–47.

"Frame Braced for Museum Space: Imperial War Museum, London." *New Civil Engineer* (8 October 1987): pp. 18, 20.

"Imperial War Museum." *Architects' Journal* (179: 1/2, 1984): p. 33.

"Imperial War Museum Phase 2." *Architects' Journal* (2 February 1994).

"Imperial War Museum." *Building Design* (15 July 1983): p. 5.

"Museums and Galleries." *Financial Times* (17 June 1989): pp. viii–ix.

"A National Heritage in Bedlam." *New Scientist* (19 January 1984).

Ostler, T. "Museums in the Air: Dr Alan Borg, the Director, Talks about the Imperial War Museum." *World Architecture* (no. 17, 1992): pp. 76–78.

Pawley, M. "Design War." *Architects' Journal* (16 August 1989): pp. 22–24.

Pawley, M. "Power Stations." *New Statesman and Society* (12 April 1991): pp. 25–26.

Docklands Light Railway Guidelines

"Docklands' Guiding Light." *Architects' Journal* (vol. 180, no. 32, 1984): pp. 22–29.

"Arup Associates' Train for Dockland Line." *Building Design* (6 April 1984): p. 3.

Stockley Park

Buttery, H. "Crafts' Untapped Talent: Stockley Park." *Designers' Journal* (April 1987): pp. 36–41.

Davey, P. "Stockley Park." *Architectural Review* (September 1989): pp. 42–48.

"Development Economics." *Architects' Journal* (27 May 1987): pp. 63–67.

Ede, B. "The Stockley Park Project." *Landscape Design* (February 1990): pp. 42–47.

Hannay, P. "Parking Business: Stockley Park." *Architects' Journal* (25 July 1990): pp. 30–43, 45–47.

Hodgkinson, P. "Mind Over Machine: Stockley Park." *Architects' Journal* (30 October 1991): pp. 44–55, 63–65.

Lyall, S. *Designing the New Landscape.* London: Thames and Hudson, 1991.

Pearman, H. "Business Parks." *Architectural Review* (March 1991): pp. 61–75.

"Potentials of Engineering." *AT* (September 1989): pp. 15–34.

"The Rebirth of the Garden." *Architectural Review* (September 1989).

Nedderhut-Heeschen, W. "B8 – Burogebaude im Stockley Park, London." *DBZ* (February 1991): pp. 217–224.

Wang, V. "Stockley Park, Londra." *Domus* (Milan, February 1990): pp. 29–37.

Legal & General House, Kingswood, Surrey

Davey, P. "Classic Study." *Architectural Review* (February 1993): pp. 37–45.

"RIBA Regional Awards." *RIBA Journal* (January 1993): pp. 31–40.

Broadgate, London

"Britain Leads Office Revolution." *Energy Management* (May 1989): pp. 16–21.

"Broadgate." *Bauwelt* (Berlin, 4 December 1992): pp. 2604–2611.

Barker, P. "Is This What We Want? Architecture in London." *Country Life* (6 April 1989): pp. 116–117.

Duffy, F. & Hannay, P. (eds). *The Changing Workplace.* London: Phaidon Press, 1992, pp. 226–236.

George, T., et al "The Design and Construction of Broadgate, Phases 1–4." *Structural Engineer* (March 1992): pp. 104–113.

Hannay, P. "Big Bang Broadgate." *Baumeister* (Munich, November 1991): pp. 19–25.

Hannay, P. "Squaring up to Broadgate." *Architects' Journal* (25 September 1985): pp. 28–31.

Herron, R. "Buildings of Distinction." *RIBA Journal* (February 1992): pp. 28–30.

Jencks, C. "Post-Modern Triumphs in London" *Architectural Design Profile* (no. 61, 1991): pp. 46–47.

Rabenek, A. "Broadgate and the Beaux Arts." *Architects' Journal* (October 1990): pp. 36–51, 57–59.

Reina, P. & Tucchman, J. "Riding on the Crest of the London Wave." *Engineering News Record* (8 December 1988): pp. 28–32.

"RIBA Awards for Architecture." *RIBA Journal* (January 1992): pp. 19–53.

"Rocks of the World." *Architecture Today* (November 1990) pp. 76, 78.

Sharp, D. "City as Theatre." *Architectural Review* (June 1989): pp. 58–65.

Smart, G. "A Civilising Influence." *Landscape Design* (May 1994): pp. 37–38.

Stocchi, A. "Il fuori Dentro." ("Unitary Planning.") *l'Arca* (Milan, April 1993): pp. 60–65.

"Stretching City Limits: A Look at the First Four Phases of Broadgate." *Building* (14 October 1988): pp. 41–48.

Bibliography Continued

Offices for Lloyds Bank, Bristol

Ashley, S. "Dockside Style: Lloyds Bank, Bristol." *Building Services* (February 1991): pp. 16–19.

"Bank Headquarters that Makes a Statement." *Building Design* (14 June 1991): pp. 14–15.

"Concrete Society Awards 1993." *Concrete* (September/October 1993): pp. 7–13.

Davies, C. "Banking on Classicism: Lloyds Bank, Bristol." *Architects' Journal* (16 October 1991): pp. 32–45, 51–53.

"Lloyds Bank Phase 2." *Building* (4 March 1994): p. 10.

Moore, R. "Bristol Fashion: Lloyds Bank in Bristol." *Foundation* (no. 3, 1992): pp. 24–27.

"Post-Modernism on Trial." *Architectural Design* (no. 11/12, 1990).

Offices for Royal Life, Peterborough

"Case Study: Fire Protection for Life." *AJ Focus* (April 1992): pp. 15–19.

Davey, P. "Insurance Premium." *Architectural Review* (May 1992): pp. 44–53.

Hannay, P. "Business Differences." *Architects' Journal* (15 July 1992): pp. 24–33.

Lyall, S. "Textbook Structure." *Concrete Quarterly* (Spring 1992): pp. 19–21.

Rognoni, M. "Offices for Royal Life, Peterborough." *Habitato Ufficio* (Milan, June 1993): pp. 50–55.

Sussex Grandstand, Goodwood

Bennetts, R. "Rational and Romantic: Arup Associates at Goodwood." *Architecture Today* (September 1990): pp. 66–73.

"Ein Leichter, Luftiger Platz. Tribune der Pferderennbahn Goodwood." *Architektur Aktuell* (October 1990): pp. 98–100.

"Roof, Sussex Grandstand." *Architects' Journal* (26 February 1992): pp. 36–39.

Paternoster Square

"A Vision for London." *Architects' Journal* (14 March 1990): pp. 28–87.

Buchanan, P. "Paternoster: Planning and the Prince." *Architects' Journal* (20 January 1988): pp. 26–29.

Buchanan, P. "Paternoster Pressure: Arup Associates' Plan for Paternoster Square." *Architectural Review* (May 1989): pp. 76–80.

Buchanan, P. "What City? A Plea for Place in the Public Realm." *Architectural Review* (November 1988): pp. 30–41.

Cruickshank, D. & Morris, N. "The Unveiling of Paternoster." *Architects' Journal* (29 May 1991): pp. 10–11.

"Fresh Thoughts on St Paul's." *Architects' Journal* (16–23 December 1987): pp. 18–19.

Magendie, F. "Londres, Tout Change!" *Techniques et Architecture* (March 1989): pp. 20–27.

"The Paternoster Redevelopment Plan, City of London." *Planning* (March 1990): pp. 22–29.

"Paternoster Square." *Techniques et Architecture* (March 1989): pp. 26–27.

"Public Design." *Architects' Journal* (6 July 1988): pp. 24–25.

Simpson, J. "Paternoster Square Redevelopment Project." *Architectural Design* (September/October 1988): pp. 78–80.

Stamp, G. "By a City Churchyard." *Spectator* (22 August 1987): pp. 13–15.

"Unbuilt London." *Architectural Review* (January 1988): pp. 14–64.

Copthorne Hotel, Newcastle

"Case Study: Windows on the Waterfront." *AJ Focus* (October 1991): pp. 15–19.

Morrison, G. "On the Waterfront: Hotel in Newcastle." *Architecture Today* (November 1991): pp. 24–26, 29.

Waste-to-Energy Plant, Belvedere, Kent

"Design for Waste-to-Energy Plant on the Thames Estuary." *Building Design* (11 June 1993): p. 8.

Fender, J. "Belvedere: To Be or Not to Be?" *Energy World* (December 1992): pp. 11–13.

Ellis Park Stadium, Johannesburg

Singmaster, D. "A Sports Stadium Designed for Protection from the Sun." *Architects' Journal* (16 March 1994): pp. 20–21.

Manchester 2000 Olympic Stadium

Binney, M. "Temples of the Sport Gods." *The Times* (14 September 1993).

Hetherington, P. (ed.). "Olympics 2000." *The Guardian* (Special supplement, 6 September 1993).

Roskrow, B. "AMEC shows the Olympic way." *Construction News* (12 August 1993): pp. 18–19.

Rowbottom, M. "Manchester's Millenium Stadium." *The Independent* (14 September 1993): p. 34.

Sudjic, D. (ed.) "Manchester: Towards 2000. The Urban Olympics." *Blueprint* (Special supplement, July/August 1993): pp. 17–21.

Sudjic, D. "Everything to Play for." *The Guardian* (3 August 1993): p. 4.

Istanbul Cultural Centre

Eczacibasi, Dr N.F. "Istanbul Culture and Congress Centre." *Tasarim* (Istanbul, August 1993): pp. 48–55.

Acknowledgments

致　　谢

The projects which are the reason for preparing this book have been made possible by the efforts of a great many people. Be they clients, architects, engineers, cost estimators, interior designers, consultants, manufacturers or builders, all have participated actively and enthusiastically as designers. They are too numerous to mention, yet without them neither book nor buildings would have been realisable.

Architecture requires patronage and our clients have been some of our most energetic supporters and inspiring collaborators. Thanks must go to all of them for their commitment, enthusiasm and support which has generated ideas and made it possible to transform those ideas into reality. We especially appreciate the efforts of Vincent Wang, Dr Alan Borg, Peter Wingrave and Dr Richard Gooder who all wrote of their experiences of working with Arup Associates.

Special thanks must go to Kathryn Taylor who, with Brian Carter, researched and prepared much of the material for this book in London. Kathryn also worked energetically with Anya Parker and Helen Allen in searching out illustrative material and preparing it for publication. The drawings were the work of many different hands, but Jirka Konopicky helped to prepare, edit and assemble that material. Thanks must also go to Denis O'Kelly and Stuart Nutton who helped enormously with the compilation of material.

The book includes the work of many different photographers. Their contribution has been invaluable and their particular views of the many buildings are incisive and inspired. We would like to thank all of them and especially Harry Sowden, Peter Mackinven, John Donat, Peter Cook, Crispin Boyle, Martin Charles, Hélène Binet and Jonathan Moore for their considerable efforts over the years.

We are grateful to Paul Latham and Alessina Brooks for their invitation to prepare this monograph and for their effort and advice in the production of it.

However, it is the many, many people of all ages, skills and interests who have worked with Arup Associates over the last thirty-one years who deserve our very special thanks. It is only as a result of their commitment to both the idea and the ideal of a way of working and creating architecture that this book has been made possible.

本书中提到的众多工程项目都是在许多人的共同努力下完成的，包括建筑用户、建筑师、工程师、预算分析师、室内设计师、顾问、材料供应商和施工人员等等，他们都积极而主动的贡献了自己的力量。由于人数众多，此处无法一一提及，但是没有他们，书中的工程项目和这本书也是不可能得以实现的。

建筑艺术需要支持和理解，我们的用户就给予了我们最有力的支持和激动人心的合作。因此，必须感谢他们，感谢他们赋予我们的重托，给予我们灵感和支持，使得我们能够将梦想变为现实。我们特别要感谢文森特·Wang、艾伦·博格博士、彼得·温格瑞夫和理查德·古德在本书的开头为我们讲述了他们与阿鲁普联合事务所合作的经验。

我们还要特别感谢凯瑟琳·泰勒和布赖恩·卡特，他们共同合作在伦敦为本书收集了大量的材料。凯瑟琳还和安亚·帕克、海伦·艾伦等一道，积极搜集本书的图片材料。此外，本书中的图片虽然出自众多设计师之手，然而是杰卡·科诺比奇帮助完成了图片的整理、编辑和排版工作。最后，还要感谢丹尼斯·欧凯利和斯图尔特·那顿在材料编辑方面的大力协助。

本书包括了众多不同摄影师的作品。他们用自己特别的视角深刻地挖掘了建筑的特色，他们的贡献是无价的。我们也要向他们致以诚挚的谢意，尤其是哈里·苏登、彼得·马克因温、约翰·多纳特、彼得·库克、克里斯平·博伊尔、马丁·查尔斯、海伦·比奈特和乔纳森·摩尔等等，他们这些年来的辛勤工作为本书添色不少。

Photography Credits

Arup Associates: 13; 15; 17; 18; 22 (2); 24 (5); 25 (6, 8); 27 (5); 28 (6, 7); 29 (8); 32 (4); 34 (7, 8); 35 (9); 40 (6); 41 (7, 9); 42 (1–3); 43 (4); 49 (3–6); 50 (1); 51 (2–4); 52 (1); 53 (2); 56 (2); 58 (6); 63 (3–5); 64 (6); 67 (10); 68 (11, 12); 69 (13); 70 (1); 71 (4); 72 (6); 74 (2); 77 (9); 84 (2); 92 (1); 93 (2); 104 (2); 108 (2); 109 (3); 125 (5, 6); 127 (3); 129 (3–5); 135 (7, 8); 138 (6); 143 (11); 154 (10, 11); 155 (12–14); 161 (6, 7); 173 (10); 177 (3, 4); 187 (3, 4); 188 (6, 8)

Roger Ball: 169 (3); 171 (6)

Hélène Binet: 173 (11)

Crispin Boyle: 76 (6); 77 (8); 79 (4); 80 (7–10); 81 (11); 89 (3, 9); 90 (4, 5); 91 (6); 93 (3); 95 (6); 96 (7); 120 (1); 121 (3, 4); 142 (10); 144 (12); 156 (2); 164 (4); 165 (5, 6); 178 (2); 179 (3); 187 (5)

Judy Cass: 110 (4); 111 (6); 112 (7)

Martin Charles: 33 (6); 75 (4); 84 (5); 85 (3–5); 86 (4, 5); 87 (7, 8); 88 (2); 122 (2); 125 (7)

Jeremy Cockayne: 176 (2); 177 (5)

John Constable: 94 (4, 5); 97 (8)

Peter Cook: Front cover; 45 (7); 46 (8, 9); 47 (10); 79 (5); 98 (2); 99 (3); 100 (4); 101 (6, 7); 102 (8); 103 (9); 105 (3, 4); 106 (5); 107 (6); 114 (2, 3); 118 (8, 9); 119 (10); 123 (3); 127 (2); 132 (1, 2); 134 (6); 137 (4); 139 (7); 146 (1–3); 147 (4–6); 156 (1); 157 (3); 161 (8); 169 (2); 171 (5); 172 (8, 9); 174 (13, 15); 175 (16); 80 (4); 181 (5, 6)

Davenport Associates: 126 (1)

Hayes Davidson: 216 (6, 7)

Richard Davis: 209 (5)

John Donat: 26 (3); 65 (7); 70 (2); 71 (3); 72 (5); 73 (7, 9)

Sir P. Dowson: 160 (1)

Richard Einzig: 38 (2); 39 (3); 58 (4)

Nicholas Gentilli: 149 (2); 150 (3)

Martin Jones: 153 (9)

Trevor Jones: 160 (2)

Peter Mackinven: 127 (4); 133 (4); 137 (5); 141 (9); 145 (13); 155 (15); Back cover.

Jonathan Moore: 173 (12); 174 (14)

Stephen Outlaw: 44 (6)

George Perkin: 124 (4)

Prudence Cuming Associates: 113 (8)

Sealand Aerial Photography: 168 (1)

Rupert Truman: 171 (7)

Trevor Walker: 57 (3); 58 (5); 61 (9)

Colin Westwood: 25 (7); 30 (1); 31 (3); 32 (5); 40 (5); 41 (8)

Illustration Credits

Ben Johnson: 113 (8); 82 background

Crispin Wride: 148 (1)

Index 索 引

Addenbrookes Site Development Plan, Cambridge, Cambridgeshire 229

Adhesive Factory, Stafford, Staffordshire 228

Al Morgab Car Park and Commercial Centre, Saudi Arabia 232

Araldite Plant, Duxford, Cambridgeshire 228

Arena, Bergiselstadion, Innsbruck, Austria **166**, 234

The Arena, Stockley Park, Uxbridge, Middlesex **156**, 233

Aston University Development Plan, Birmingham 230

Bab Al Sheikh, City Centre Housing and Commercial Development, Baghdad, Iraq 232

Babergh District Council Offices, Hadleigh, Suffolk **92**, 232

Bedford High School, Bedford, Bedfordshire 233

Bedford School, Bedford, Bedfordshire **88**, 232

Berlin 2000 Olympic Facilities, Berlin, Germany **190**, 235

Boringdon Park, Plymouth, Devon 234

Boulton House, Trinity Hall, Cambridge, Cambridgeshire 229

Briarcliff House, Farnborough, Hampshire **104**, 232

Bridewell Development, London 235

Broadgate, London **126**, 233

Bus Station and Offices, Northampton, Northamptonshire 231

Bush Lane House, Cannon Street, London **52**, 231

Canning Factory, Melton Mowbray, Leicestershire 231

Canteen Building, Peterborough, Cambridgeshire 231

CEGB South West Region Headquarters, Bedminster Down, Bristol **78**, 231

Central Offices, Peterborough, Cambridgeshire 230

Chemical Plant, Newton Aycliffe, Durham 232

Chemical Plant, Tonbridge, Kent 228

Civil Engineering Department, Loughborough, Leicestershire 229

Computer Building, Sheffield, South Yorkshire 231

Computer Centre, Phase III, Cosham, Hampshire 231

Copthorne Hotel, Newcastle-upon-Tyne **176**, 234

Department of Arts and Commerce, Muirhead Tower, Birmingham 229

Department of Arts and Social Sciences, Attenborough Tower, Leicester, Leicestershire 229

Department of Chemical Engineering, Loughborough, Leicestershire 229

Department of Metallurgy, Oxford, Oxfordshire 231

Depot and Housing, Warwick Road, London 231

Diplomatic Quarter Police Headquarters, Riyadh, Saudi Arabia 232

Diplomatic Quarter Sports Club, Riyadh, Saudi Arabia **120**, 232

Docklands Light Railway Design Guide, London 232

Düsseldorf Tower, Düsseldorf, Germany **200**, 235

Electron Microscope Building, Birmingham 230

Electron Microscope Building, Oxford, Oxfordshire 229

Eli Lilly Research Centre, Windlesham, Surrey 235

Ellis Park Athletics Stadium, Johannesburg, South Africa 235

Factory for Trebor Limited, Colchester, Essex **84**, 232

Factory for York Shipley Limited, Basildon, Essex 228

Factory, Ashford, Kent 228

Factory, Hemel Hempstead, Hertfordshire 228

Factory, Laboratory and Offices, Welwyn Garden City 228

Factory, London 228

Finished Goods Store, Melton Mowbray, Leicestershire 232

Forbes Mellon Library, Clare College, Cambridge, Cambridgeshire **122**, 233

Formalin Plant, Duxford, Cambridgeshire 228

FPW Markets, Istanbul, Turkey 235

Fry Building, Somerville College, Oxford, Oxfordshire 229

Gateway 1, Basingstoke, Hampshire **56**, 231

Gateway 2, Basingstoke, Hampshire **98**, 232

Gonville and Caius College, Cambridge, Cambridgeshire **192**, 235

Government Office, Great George Street, London 235

Government Offices Study, 2 Marsham Street, London 234

Grand Lyon Porte des Alpes, Lyon, France **206**, 235

Great Common Farm, Bourn Airfield, Cambridgeshire 234

Gro, Newtown, Wales **212**, 235

Hammamasu Resort, Hokkaido, Japan 234

Hasbro Bradley (UK) Limited Headquarters, Stockley Park, Uxbridge, Middlesex **146**, 233

Henry Wood Hall, Southwark, London **48**, 231

Hillat al Hammad Car Park and Commercial Centre, Saudi Arabia 232

HM Naval Base Dock Support Building, Portsmouth, Hampshire 231

HM Naval Base Heavy Plate Shop, Portsmouth, Hampshire 230

Hong Kong Central Station, Lantau and Airport Railway, Hong Kong **196**, 235

Honshu Paper, Fukuoka, Japan 235

Horizon Project, Factory for John Player & Sons, Nottingham **38**, 230

Horsham Park, Genshagen, Germany **182**, 234

House, Iver, Buckinghamshire 231

IBM (UK) Limited Computer Centre and Assembly Plant, Havant, Hampshire **30**, 229

IBM Building, Johannesburg, South Africa **50**, 229

IBM North Harbour, Portsmouth, Hampshire **42**, 229

The Imperial War Museum, Stage I, Lambeth, London **148**, 233

The Imperial War Museum, Stage II, Lambeth, London 235

Industrial Research Laboratories, Horsham, Surrey 228

Inmarsat Place, London 234

International Garden Festival Hall, Liverpool **108**, 232

Istanbul Cultural and Congress Centre, Istanbul, Turkey **208**, 235

Johannesburg Athletics Stadium, Johannesburg, South Africa **218**, 235

Jubilee Line Extension Service Control Centre, Neasden, London **184**, 234

Leckhampton House, Corpus Christi College, Cambridge, Cambridgeshire 228

Legal & General House, Kingswood, Surrey **136**, 233

Library and Shops, Lion Yard City Centre Development, Cambridge, Cambridgeshire 231

Lion Yard City Centre Development Plan, Cambridge, Cambridgeshire 229

Lion Yard Magistrates' Court, Cambridge, Cambridgeshire 232

The Liver Building, Liverpool 232

Lloyd's of London Administrative Headquarters, Chatham, Kent **74**, 231

Lloyd's of London Press, Colchester, Essex 233

The London Coliseum, London 231

London Luton Airport, Luton, Bedfordshire 235

London Transport Bayswater Station, London 233

London Transport Embankment Station, London 233

London Transport Notting Hill Gate Station, London 233

Loughborough Master Plan, Loughborough, Leicestershire 229

Maintenance Building, Duxford, Cambridgeshire 228

The Maltings Concert Hall, Snape, Aldeburgh, Suffolk **26**, 229

Manchester 2000 International Olympic Stadium, Manchester **214**, 235

Manchester Airport Car Park, Manchester 234

Mechanical Engineering Building, Loughborough, Leicestershire 229

Microbiology Department, Birmingham 230

Milton Bradley Fitting Out Building B1, Stockley Park, Uxbridge, Middlesex 233

Mining and Metallurgy, University of Birmingham, Birmingham **22**, 228

Multi-Purpose Building, Duxford, Cambridgeshire 228

Multi-Storey Car Park, Hull, Humberside 228

Multi-Storey Car Park, Lion Yard City Centre Development, Cambridge, Cambridgeshire 230

Music School, Norwich, East Anglia 230

National Bowling Centre, Worthing, West Sussex 235

New Museums, Cambridge, Cambridgeshire 230

Nuclear Physics, Oxford, Stage III, Oxfordshire 229

Nursery School, Sossenheim, Frankfurt, Germany **198**, 235

Office and Warehouse, Bury St Edmunds, Suffolk 230

Office Building and Commercial Centre, Dammam, Saudi Arabia 232

Office Extension, Welwyn Garden City, Hertfordshire 228

Offices and Shops, Oxford Street, London 234

Offices for Lloyds Bank PLC, Canons Marsh, Bristol **168**, 234

Offices, Harmondsworth, Greater London 230

Offices, Peterborough, Cambridgeshire 230

1 Finsbury Avenue, London **114**, 233

123 Buckingham Palace Road, London **178**, 234

Open Air Pool, Eton, Berkshire 232

Opera House Restoration, Buxton, Derbyshire 232

Oxford Railway Station, Oxford, Oxfordshire 234

Paper Mill, Jebba, Nigeria 228

Paternoster Square, London **128**, 233

Pavilion, Cowes, Isle of Wight 230

Pears Britten Music School, Snape, Suffolk 231

Physics, Electrical Engineering, Chemical Engineering Departments, Loughborough, Leicestershire 230

Plantation House, London 234

Point Royal Flats, Bracknell, Berkshire 228

Post Offices, Cape Coast 228

Printing Works and Offices, Oxford, Oxfordshire 230

Private House, Drax Avenue, Wimbledon, London 229

Projet Suroit Development Plan, South West Mauritius 230

Research Laboratories, Duxford, Cambridgeshire 228

Residential Buildings, Loughborough, Leicestershire 229

Residential Buildings, Staverton Road, Oxford, Oxfordshire 230

Royal Insurance House, Peterborough, Cambridgeshire **186**, 234

Royal Maritime Auxiliary Service Maintenance and Support Centre, Portsmouth, Hampshire 231

Saudi Arabian National Guard Housing, Dirab, Saudi Arabia 233

School Buildings, Ampleforth, North Yorkshire 231

Science Laboratories, Eton, Berkshire 232

Scotstoun House, South Queensferry, Lothian 229

Sheffield University Development Plan, Sheffield 230

Sir Thomas White Building, St John's College, Oxford, Oxfordshire **70**, 231

Skinners Hall, London 233

Skippets House, Basingstoke, Hampshire 233

Sports Centre, Ampleforth, North Yorkshire 230

Sports Hall, Guildford, Surrey 229

Stockley Park A3 Building, Stockley Park, Uxbridge, Middlesex 233

Stockley Park B5 Building, Stockley Park, Uxbridge, Middlesex 233

Stockley Park Building A3.2, Stockley Park, Uxbridge, Middlesex 233

Stockley Park, Heathrow, Uxbridge, Middlesex **132**, 233

Sussex Grandstand, Goodwood, Sussex **160**, 234

Swimming Pool, Eton, Berkshire 232

Swimming Pool, Walton-on-Thames, Surrey 228

Systems Assembly Building and Offices, Reading, Berkshire 232

Teaching Building, Guildford, Surrey 230

Technology Park, Buda Nova, Hungary 234

Telephone Exchange, Accra North, Ghana 229

Theatre Royal Restoration, Glasgow 231

Triton Square, London 234

Truman Limited Headquarters, Brick Lane, London **62**, 231

Varnish Kitchen, Stafford, Staffordshire 228

Vaughan Building, Oxford, Oxfordshire 230

Vik Supplies Factory, Stafford, Staffordshire 228

Villebon sur Yvette, Orly, France 234

Warehouse and Book Store, Harmondsworth, Greater London 229

Waste-to-Energy Plant, Belvedere, Kent **204**, 235

Watling House, London 234

Wentworth Golf Club, Wentworth, Surrey **162**, 234

Wimbledon Bridge House, Wimbledon, London 233

Wolfson Building, Somerville College, Oxford, Oxfordshire 229

Wychfield House, Trinity Hall, Cambridge, Cambridgeshire 230

XXth District Park, Budapest, Hungary 234

Zimbabwe House Renovation, The Strand, London 232

Every effort has been made to trace the original source of copyright material contained in this book. The publishers would be pleased to hear from copyright holders to rectify any errors or omissions.

The information and illustrations in this publication have been prepared and supplied by Arup Associates. While all reasonable efforts have been made to ensure accuracy, the publishers do not, under any circumstances, accept responsibility for errors, omissions and representations express or implied.

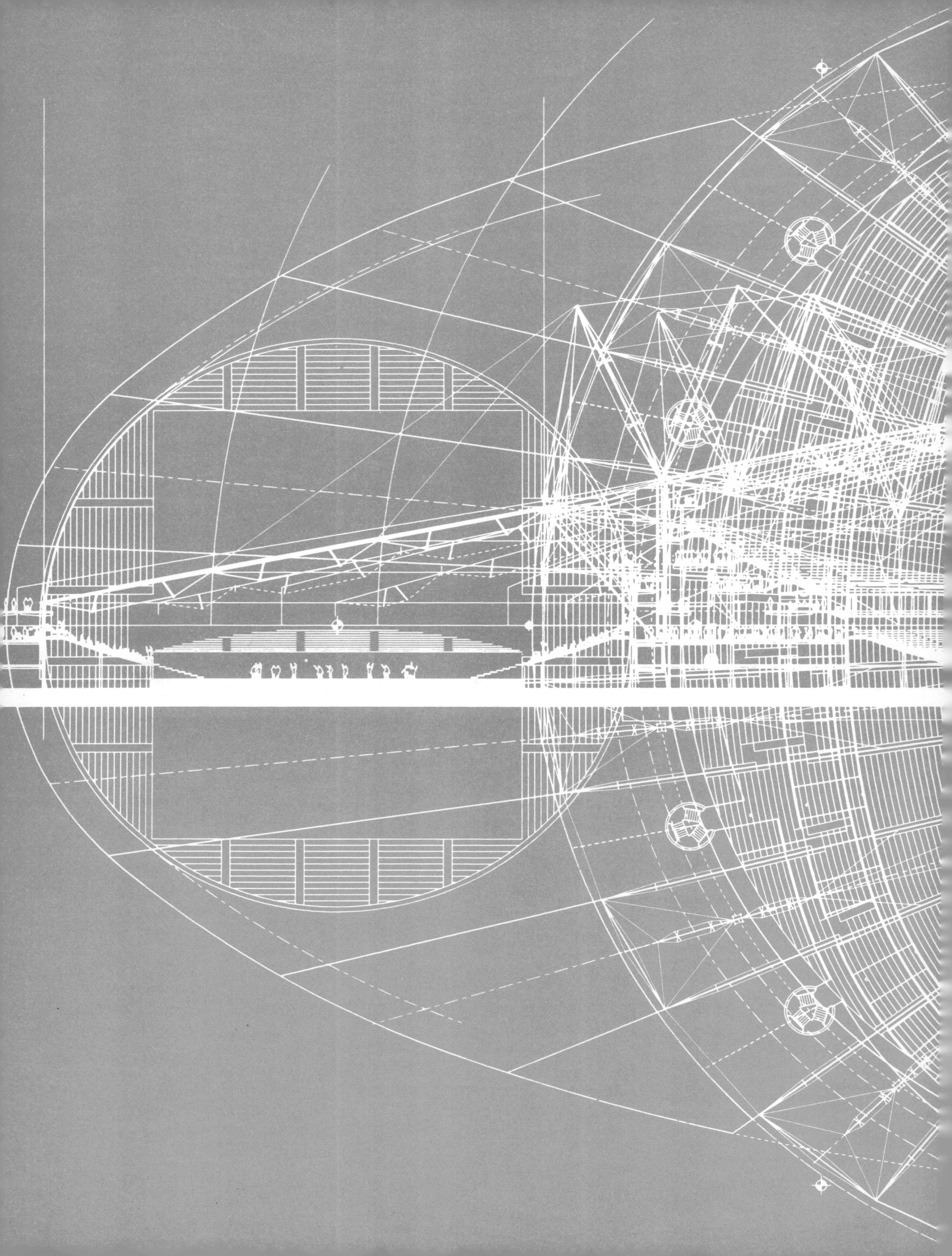